Advances in Anatomy Embryology and Cell Biology

Vol. 151

Springer-Verlag Berlin Heidelberg GmbH

Birte Steiniger Peter Barth

Microanatomy and Function of the Spleen

With 10 Figures and 2 Tables

Springer

Birte Steiniger
Institut für Anatomie und Zellbiologie
Robert-Koch-Straße 6
D-35033 Marburg, Germany
(e-mail: steinigb@post.med.uni-marburg.de)

Peter Barth
Medizinisches Zentrum für Pathologie
Baldingerstraße
D-35043 Marburg, Germany

Library of Congress-Cataloging-in-Publication-Data
Steiniger, Birte, 1955- . Microanatomy and function of the spleen / Birte Steiniger, Peter Barth. p. cm. - (Advances in anatomy, embryology, and cell biology; v. 151)
Includes bibliographical references and index.
ISBN 978-3-540-66161-0 ISBN 978-3-642-57088-9 (eBook)
DOI 10.1007/978-3-642-57088-9
M.D. II. Title. III. Series. [DNLM: 1. Spleen-anatomy & histology. 2. Spleen-physiology. W1 AD433K v. 151 2000 / WH 600 S822m 1999] QM576.S84 1999 612.4'1-dc21 DNLM/DLC for Library of Congress

Production: PRO EDIT GmbH, D-69126 Heidelberg
SPIN: 10698805 27/3136-5 4 3 2 1 0 - Printed on acid-free paper

Contents

Abbreviations

ABC	Avidin-biotinylated peroxidase complex
APAAP	Alkaline-anti-alkaline phosphatase complex
BSA	Bovine serum albumin
CA	Central arteriole
CD	Cluster of differentiation
CO	Corona
DAB	3,3′-Diaminobenzidine
FDC	Follicular dendritic cell
GC	Germinal centre
HE	Hematoxilin-eosin
IDC	Interdigitating dendritic cell
Ig	Immunoglobulin
iMZ	Inner marginal zone
LCA	Leukocyte common antigen
MAb	Monoclonal antibody
MHC	Major histocompatibility complex
MMMϕ	Marginal metallophilic macrophages
MF	Myofibroblasts
MS	Marginal sinus
MZ	Marginal zone
MZMϕ	Marginal zone macrophages
NK cells	Natural killer cells
oMZ	Outer marginal zone
oPALS	Outer periarteriolar lymphatic sheath
PBS	Phosphate-buffered saline
PFZ	Perifollicular zone
PALS	Periarteriolar lymphatic sheath
RT	Room temperature
SC	Sheathed capillary
SCID	Severe combined immunodeficiency
SPF	Specified pathogen-free
TCR	T cell receptor
TI antigens	T cell-independent antigens
TI-2 antigens	T cell-independent type 2 antigens
TNF	Tumour necrosis factor

1 Introduction

The spleen remains one of the few human organs with a still enigmatic microanatomical structure and an obscure function. Rather detailed and precise microscopic investigations on the human spleen were carried out in the late nineteenth and at the beginning of the twentieth centuries (Schweigger-Seidel 1863; Weidenreich 1901; Jäger 1929). However, since then increasing confusion about the microscopic structure of the organ has arisen, because several authors have uncritically amalgamated findings in different animal species with those in man. Thus, present textbooks of histology offer an astonishing variety of models depicting human splenic microarchitecture. There is, for example, no consensus with respect to the distribution of T and B lymphocytes, macrophages and stromal cells. In addition, the presentation of human splenic microvasculature is also quite diverse and a consistent nomenclature for different parts of the vascular tree is lacking. This situation is not only typical of textbooks, but it is also reflected in the primary literature. Thus, discrepant findings concern the number and arrangement of distinguishable regions in human splenic secondary follicles or the existence of a marginal zone (MZ) around T cell regions. With respect to the microvasculature, the location and function of "penicillar arterioles", "sheathed capillaries" or the "marginal sinus" and "perimarginal sinus" of the white pulp are far from clear.

This review is intended to provide answers to some of the controversial issues of human splenic microanatomy and function. As the answers stand out more clearly in the form of a comparison, we contrast the microstructures of human and rat spleen, beginning with the latter species. This is necessary, because up to now splenic structure and function, as well as lymphocyte migration through the spleen, have been most thoroughly investigated in rats. Furthermore, the nomenclature for certain parts of the splenic microvasculature and for special splenic macrophage populations was coined in rats.

Thus, in Chap. 3, we introduce the microscopic anatomy of the spleen by first providing an overview of the current knowledge of splenic microanatomical compartments in rats. Then, in Chap. 4, we focus on the species-specific structures observed in human spleen. The human spleen has always been a rather neglected organ and only a few detailed immunohistological studies are available which have investigated its microanatomical regions and their cell populations. The most problematic region of the human spleen is the border between white and red pulp, which serves as an entry compartment for antigens and recirculating lymphocytes, but also is the habitat of a special type of B memory cells. The emphasis in Chaps. 3 and 4 is on this region, summarising results mainly obtained in our laboratory (Steiniger et al. 1993, 1997). Chapter 4 contains new and partially unpublished findings made on human spleens.

We utilise the nomenclature and definitions established in rats to show that the border between red and white pulp has a more complicated structure in humans than in rodents. Chapters 5–7 then review the current state of knowledge on the immunological function of different splenic compartments. For obvious reasons, this knowledge is almost entirely based on investigations performed in rodents. It remains open to speculation as to how far the results may be extrapolated to man. Finally, Chap. 8 covers open questions of immunological and physiological relevance associated with human spleens. At the end of Chap. 8, a hypothesis about lymphocyte migration through the human splenic white pulp is presented.

In contrast to inbred laboratory rats, humans exhibit considerable inter-individual variability of splenic microanatomy. Thus, the number of human spleens examined is of crucial importance to provide precise immunohistological data before tackling more sophisticated questions. The scope of this review is to outline the basic and more or less invariant immunohistological features of the organ. We have primarily investigated spleens removed from young, severely polytraumatised patients, which we do not expect to over-represent haematologic malignancies or other abnormalities. An analysis of the multiple factors that may be responsible for the variability of lymphatic tissue, such as age, "outbred" genetic background, environmental antigens and pathogens is not within the scope of this publication. Finally, the clinical situation before removal of the organ and medical intervention per se are not considered, although both certainly influence human spleen differently from that of laboratory rats.

The following well-established overall scheme of splenic architecture in rodents forms the basis of our investigation. The parenchyma of the spleen consists of two compartments, the white pulp and the red pulp. The white pulp is arranged more or less concentrically around the so-called "central arterioles." It represents the part of the organ with the highest density of lymphocytes. The red pulp contains most of the erythrocytes due to its special structure, which is composed of two types of blood-filled spaces, the sinuses (also called sinusoids) and the splenic cords. The splenic cords represent a meshwork of connective tissue inhabited by lymphocytes, macrophages, plasma cells and haematopoietic cells. The white pulp of the spleen comprises three sub-compartments, the periarteriolar lymphatic sheath (PALS), the follicles and the MZ. The PALS represents the region directly surrounding the central arterioles, where T lymphocytes predominate. Follicles are aggregations of migratory B lymphocytes attached at the periphery of the PALS. They exist either as primary follicles of uniform internal structure or as secondary follicles containing a pale-staining germinal centre. The MZ surrounds both PALS and follicles and delimits these compartments from the red pulp. The majority of lymphocytes in the MZ normally belong to a special type of memory B lymphocytes. Some authors regard the MZ as a region of its own in addition to the white and red pulp. This distinction is, however, only a matter of semantics.

2 Materials and Methods

2.1 Animal Spleens

The spleens of 40 male LEW/Ztl rats, bred under specified pathogen-free (SPF) conditions in the Zentrales Tierlabor of Hannover Medical School, were investigated, when the animals weighed 300 g at an age of about 3 months. The rats were killed by ether anaesthesia and organ specimens were either fixed immediately in Bouin's solution for 24 h before paraffin embedding, or they were frozen and stored in liquid nitrogen for cryostat histology. In addition, a total of ten LEW rats of the same age were housed together with other animals of unknown derivation under "conventional" conditions without any special hygienic precautions.

2.2 Human Spleens

Specimens from 44 individuals (23 males and 21 females) were investigated. Of these, 26 specimens came from individuals aged under 35 years. Twenty-eight specimens were taken from organs removed because of traumatic rupture caused by major accidents. In five cases, a splenectomy was performed in combination with a gastrectomy because of gastric malignancy. In two cases, a widespread colonic carcinoma led to splenectomy. One specimen came from a patient with pancreatitis. In eight cases, the specimens were obtained post-mortem from patients with other malignancies (5), cardiovascular diseases (2) and CO intoxication (1). The organs had been either fixed in 4% buffered formalin (n=32) for an unknown length of time or they had been kept in Baker's formol-calcium for 24 h (n=12) and were immediately embedded in paraffin. In five cases, the specimens were frozen and stored in liquid nitrogen within 4 h after removal of the organ.

2.3 Antibodies

The primary antibodies used to detect human and rat antigens are summarised in Tables 1 and 2.

Polyclonal and monoclonal anti-human sialoadhesin antibodies were kindly donated by Dr. P.R. Crocker and Dr. A. Hartnell, Institute of Molecular Medicine,

Table 1. Antibodies against human antigens

MAb	Specificity	Producer/Distributor	Reference
28	Human CD15, 3-fucosyl-N-acetyl-lactosamine (Le^x) residue, present in several membrane molecules on monocytes and granulocytes	Cymbus Biotechnology, Chandlers Ford, UK, No. CBL 144, via Dianova, Hamburg, Germany	
1F6	Human CD4	Novocastra, Edinburgh, UK, No. NCL CD4/CD8, via Medac, Hamburg, Germany	
4B11	Human CD8, alpha chain	Novocastra, Edinburgh, UK, No. NCL CD4/CD8, via Medac, Hamburg, Germany	
asm-1	Smooth muscle alpha-actin	Progen GmbH, Heidelberg, No. 61001	Skalli et al. 1986
CR3/43	Human MHC class II molecules (HLA-DP/DQ/DR)	DAKO, Hamburg, Germany, No. M 0775	
	Human sialoadhesin, polyclonal rabbit IgG	Donated by Dr. P.R. Crocker/ Dr. A. Hartnell	Unpublished
HSN1	Human sialoadhesin, MAb	Donated by Dr. P.R. Crocker/ Dr. A. Hartnell	Unpublished
	Human IgM, polyclonal rabbit Ig	DAKO, Hamburg, Germany, No. A 0425	
	Human IgD, polyclonal rabbit Ig	DAKO, Hamburg, Germany, No. A 0093	
	Human CD3 epsilon chain (aa 156–168), polyclonal rabbit Ig	DAKO, Hamburg, Germany, No. A 0452	
JC/70 A	Human CD31 (PECAM-1), expressed on endothelia and other cells	DAKO, Hamburg, Germany, No. M 0823	Parums et al. 1990
JCB117	Human CD79a (Ig-alpha or mb-1), signal transducing molecule associated with membrane Ig	DAKO, Hamburg, Germany, No. M 7050	
KiM1-P	Human CD68 (macrosialin), expressed by most monocytes and macrophages	Donated by Dr. M.R. Parwaresch,University of Kiel, Germany	Radzun et al. 1991
KiM2	Human monocytes, certain macrophages and granulocytes, antigen unknown	Donated by Dr. M.R. Parwaresch, University of Kiel, Germany	Radzun and Parwaresch 1983

Table 1. *Continued*

MAb	Specificity	Producer/Distributor	Reference
L26	Human CD20, putative Ca^{2+} channel expressed primarily on B cells	DAKO, Hamburg, Germany, No. M 0755	Mason et al. 1990
MϕP-9	Human CD14 (Leu-M3), expressed by monocytes, granulocytes (weakly) and macrophage subpopulations; acts as receptor for bacterial lipopolysaccharide	Becton Dickinson, Heidelberg, Germany, No. 7490	
MNF116	Human cytokeratins 5,6,8 and 17	DAKO, Hamburg, Germany, No. M 0821	
TM 1009	Human thrombomodulin	DAKO, Hamburg, Germany, No. M 0617	
Tu1	Human CD23, low affinity Fc-receptor for IgE (FcεRII); expressed among other cells on recirculating B lymphocytes	Cymbus Biotechnology, Chandlers Ford, UK, No. CBL 156, via Dianova, Hamburg, Germany	

Table 2. Antibodies against rat antigens

MAb	Specificity	Producer/Distributor[a]	Reference
ED 1	Rat lysosomal membrane antigen, similar to CD68 (macrosialin), expressed by most monocytes and macrophages	Serotec, Oxford, UK, No. MCA 341	Dijkstra et al. 1985a
ED 2	175 kDa membrane glycoprotein expressed by subpopulation of mature rat macrophages, not by monocytes	Serotec, Oxford, UK, No. MCA 342	Dijkstra et al. 1985a
ED 3	Most probably rat sialoadhesin, expressed by subpopulations of macrophages, not by monocytes	Serotec, Oxford, UK, No. MCA 343	Dijkstra et al. 1985a
HIS52 (RECA-1)	Antigen unknown, expressed by rat endothelial cells	Medac, Hamburg, Germany	Duijvestijn et al. 1992
KiM9R	Antigen unknown, expressed by MMMϕ, MZMϕ and by subsinusoidal and medullary macrophages of lymph nodes	Donated by , Dr. H.H. Wacker University of Kiel, Germany	Wacker 1994
Ox 3	H-2A-like MHC class II molecules, polymorphic epitope present in LEW	Serotec Oxford, UK, No. MCA 45	McMaster and Williams 1979

Table 2. *Continued*

MAb	Specificity	Producer/Distributor[a]	Reference
Ox22	Rat CD45 RC, high molecular weight forms of the leukocyte common antigen containing product of exon C, expressed by B and certain T lymphocytes	Serotec, Oxford, UK, No. MCA 53	Spickett et al. 1983
Ox33	Most probably rat CD45 RA, high molecular weight form of the leukocyte common antigen containing product of exon A, expressed only by B lymphocytes	Serotec, Oxford, UK, No. MCA 340G	Woollett et al. 1985
R73	Rat alpha/beta T cell antigen receptor, constant domain of beta-chain	Serotec, Oxford, UK, No. MCA 453G	Hünig et al. 1989

[a] The present distributor of Serotec products in Germany is Biozol, Eching, Germany.

University of Oxford, and will be reported elsewhere (P.R. Crocker et al., unpublished). Briefly, affinity-purified rabbit anti-human sialoadhesin IgG was prepared by injecting rabbits with a recombinant human sialoadhesin-Fc fusion protein, consisting of the first four N-terminal Ig-like domains of sialoadhesin linked to the Fc portion of human IgG1. The purified immune IgG was depleted of Fc reactivity on a column of human IgG-Sepharose, followed by purification of anti-sialoadhesin IgG on a sialoadhesin-Fc-Sepharose column. Mouse IgG1 anti-human sialoadhesin (MAb HSN1) was generated by immunising mice with recombinant human sialoadhesin-Fc and screening hybridoma supernatants for specific reactivity with the immunogen.

Peroxidase-conjugated rabbit anti-mouse immunoglobulin (No. Z 109) was obtained from DAKO, Hamburg, Germany. Biotinylated goat anti-rabbit IgG (No. BA-1000) and the ABC elite kit (No. PK-6102) were purchased from Vector Laboratories, Burlingame, USA, via Camon, Wiesbaden, Germany.

2.4 Single Staining Procedure for Immunohistology

Both paraffin and cryostat sections were investigated to assess the normal distribution of antigens in human spleen. This is essential because, due to formalin fixation, paraffin sections often exhibit reduced antibody reactivity in comparison to cryosections. On the other hand, the resolution of cryosections is often too low to permit unequivocal identification of splenic compartments. Thus, in Chap. 4, the type of sections used is always reported.

Most human antigens, as well as smooth muscle actin in rats, were detected in cryostat and paraffin sections using the Vectastain ABC elite kit. CD14, CD15, CD23 and KiM2 were only detected in cryostat sections, while CD79a and CD4 were only studied in paraffin sections. Paraffin sections were deparaffinized, washed in bidistilled water and incubated in 0.05% proteinase, type XIV (Sigma No. P-5147) in Tris/NaCl, pH 7.6, for 15 min at room temperature (RT). After washing in PBS, the sections were covered with H_2O_2 solution at a final concentration of 0.15% in PBS for 30 min at RT to block peroxidatic activity. Cryostat sections were fixed in cold isopropanol for 10 min and blocked with 0.05% phenylhydrazine in PBS for 20 min at RT. The primary antibodies were used at dilutions between 1:100 and 1:800 in PBS, containing 1% BSA, 0.1% NaN_3 and a final concentration of 3 µg/ml avidin (Sigma No. A-9275) and incubated with the washed sections overnight at 4°C. Biotinylated goat anti-rabbit or anti-mouse IgG was diluted 1:200 in PBS with a final concentration of 20 µg/ml biotin (Sigma No. B 4501). After washing, the sections were incubated with this solution for 30 min at RT. The avidin-biotinylated peroxidase complex was prepared according to the manufacturer's instructions 30 min before use by sequentially diluting solutions A and B in PBS to a final concentration of 1:50. The sections were washed and incubated with the avidin-biotinylated peroxidase complex for 30 min at RT. After washing, peroxidase activity was revealed using diaminobenzidine and the sections were then dehydrated and coverslipped in Eukitt. Alternatively, Vector VIP (Vector Laboratories, Burlingame, USA, No. SK-4600) was used as chromogen and the sections were coverslipped in Mowiol and photographed immediately.

MAb CR3/43 against human MHC class II molecules was visualised by a tyramide-enhanced ABC method on paraffin sections. The MAbs directed against CD79a, CD4, CD8 and against cytokeratins were used on paraffin sections which had been autoclaved on silanised slides for 20 min in citrate buffer.

If formol pigment was unacceptable, paraffin sections were deparaffinized to 80% alcohol, incubated with a saturated solution of picric acid for 30 min at RT, and washed in 80% alcohol until the yellow colour had vanished before routine deparaffinisation was completed.

ED1, ED2, ED3, Ox3, KiM9R and R73 were detected in cryostat sections of rat spleens using an indirect immunoperoxidase technique. Ox22 and Ox33 were demonstrated by the ABC or APAAP method.

2.5 Double Staining Procedure for Immunohistology

A subtractive double-staining procedure using the CD68 antibody KiM1-P and MAb HSN1 against sialoadhesin was performed on paraffin sections to prove that the sialoadhesin-positive cells were macrophages.

MAb KiM1-P was visualised first using the ABC method and diaminobenzidine (DAB), as described above. After appropriate washes, MAb HSN1 supernatant was diluted 1:10 in PBS/BSA/NaN_3 and applied for 1 h at RT. HSN1 was revealed by rabbit anti-mouse Ig (1:30 in PBS, for 30 min) and mouse APAAP-complexes (1:40 in PBS, for 30 min). Incubation with anti-Ig and APAAP-complexes was repeated once to enhance staining. Finally, deposition of APAAP complexes was revealed using fast blue BB as described in Sect. 2.7, but including levamisol in the AS-MX-phosphate buffer.

2.6
Demonstration of Acid Phosphatase in Cryostat Sections

Acid phosphatase activity was detected in cryostat sections according to the method of Burstone and Barka. Briefly, 400 mg pararosaniline-chloride (Acros Organics, New Jersey, USA, No. 41680-0250) was dissolved in 10 ml 2 N HCl and filtered. 400 mg $NaNO_2$ was dissolved in 10 ml bidistilled water. 250 µl of pararosaniline solution were added to 250 µl of $NaNO_2$ solution and allowed to react for 1–2 min. The mixture was then immediately diluted with 20 ml of Veronal-acetate buffer, pH 7.6. 10 mg naphthol-AS-biphosphate (Sigma No. 2125) was dissolved in 1 ml N,N-dimethylformamide and added to the buffered pararosaniline solution. The pH was adjusted to 5.0 and the solution was warmed to 37°C. Cryostat sections were fixed for 10 min in Baker's formol-calcium at 4°C, washed in bidistilled water and incubated for 15 min at 37°C in the prewarmed hexazotised pararosaniline/naphthol-AS-biphosphate solution. The sections were then washed with distilled water, counterstained with Mayer's hemalum and coverslipped in Mowiol.

2.7
Demonstration of Alkaline Phosphatase in Cryostat Sections

Cryostat sections were fixed in cold isopropanol for 10 min and incubated in the dark for 20 min at RT with 1 mg/ml fast blue BB salt (Sigma No. F 3378) in naphthol AS-MX-phosphate buffer. The buffer was prepared by dissolving 2 mg naphthol AS-MX-phosphate (Sigma No. N 4875) in 0.2 ml dimethylformamide (Merck No. 3034) and adding this solution to 9.8 ml of 0.1 M Tris-HCl buffer, pH 8.2. The fast blue solution was filtered before use. The sections were counterstained with nuclear fast red and mounted in Mowiol.

3 Microanatomical Compartments of the Rat Spleen

3.1 White Pulp

In normal spleens of 3-month-old inbred LEW rats kept under specified pathogen-free conditions, the white pulp occupies a substantial portion of the entire organ parenchyma. In rats reared under these conditions, the structure of the splenic white pulp is individually invariant. During immune reactions the red pulp area may, however, increase substantially because haematopoiesis, which normally occurs in adult splenic red pulp in rats, will be enhanced. As previously described by several investigators (Veerman and van Ewijk 1975; Dijkstra et al. 1985a; van Rooijen et al. 1989; Kraal 1992), the white pulp consists of three distinct compartments, the PALS, the follicles and the MZ (Figs. 1, 2A).

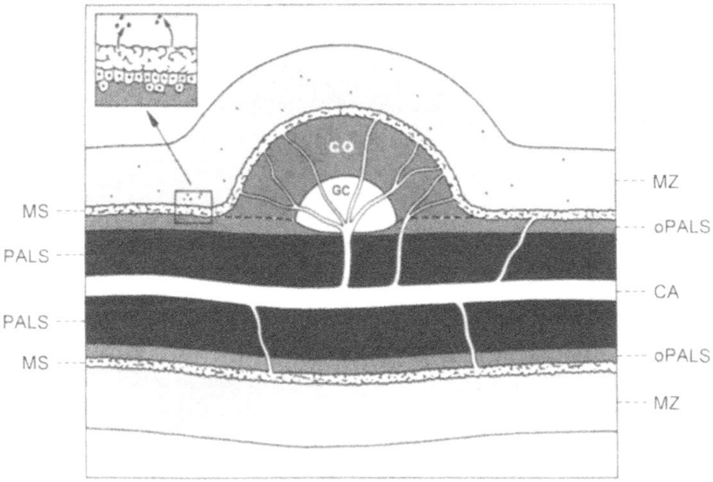

Fig. 1. Schematic drawing of a longitudinal section through the rat splenic white pulp: the central arteriole (*CA*) is accompanied by a broad periarteriolar lymphatic sheath (*PALS*) predominantly inhabited by T cells. The outer PALS (*oPALS*) also serves as a recirculation compartment for stimulated B cells. The germinal centre (*GC*), the corona (*CO*) and the marginal zone (*MZ*) are B cell regions. PALS, germinal centre and corona are separated from the marginal zone by the marginal sinus (*MS*). The *inset* demonstrates that according to Kraal (1992) the inner wall of the marginal sinus is lined by marginal metallophilic macrophages, while the outer wall is leaky and permits extravasation of particles and blood cells.

3.1.1
Periarteriolar Lymphatic Sheath

The innermost compartment of the white pulp is called the "periarteriolar lymphatic sheath" (PALS). It is located concentrically around the so-called central arterioles, which traverse the entire organ. The PALS is a compartment harbouring predominantly T lymphocytes as demonstrated immunohistologically by detection of the alpha/beta T cell antigen receptor (TCR) (Fig. 3A), the CD5 molecule and various other T lymphocyte cell-surface glycoproteins. In conventional HE-stained sections the T lymphocytes are visualised as small densely packed cells with extremely dark nuclei and minimal amounts of cytoplasm (Fig. 2A). In addition, the PALS contains specialised dendritic cells, interdigitating dendritic cells (IDC), the nuclei of which stain more lightly. These cells express high amounts of cell-surface class II major histocompatibility complex (MHC) antigens and function as antigen-presenting cells for T lymphocytes (Fig. 3B). In addition, they are positive for ED1, which is supposed to detect CD68 (Fig. 4A). Few B lymphocytes are present in the PALS, especially in its outer part. The outer PALS (oPALS) is a compartment of initial antigen-specific interaction among T lymphocytes and migrating B lymphocytes during primary and secondary immune reactions (see Sect. 5.1.2). These migrating B cells form a thin chain or sleeve of cells at the periphery of the PALS. They are in continuity with primary follicles or the corona of secondary follicles (Fig. 2C, D). A rather ill-defined population of macrophages in the oPALS stains with the MAb ED2 (Fig. 4C) and may assist B cell differentiation (Matsuno et al. 1989). ED2 is a cell surface antigen of unknown function, which is present on a subpopulation of mature macrophages (Dijkstra et al. 1985). The inner scaffold of the PALS is formed by a special type of fibroblast or "reticulum cells" expressing smooth muscle alpha actin, as shown by MAb asm-1, and alkaline phosphatase. These cells form a fine meshwork which is especially prominent in the oPALS close to the marginal sinus, while the inner PALS is supported by coarser meshes (Fig. 3C). Several monoclonal antibodies have been developed against rat reticulum cells, which also recognise these fibroblasts (van den Berg et al. 1989).

3.1.2
Follicles

In rats kept under specified pathogen-free conditions more or less hemispherical accumulations of B lymphocytes, so-called primary follicles, are attached to the PALS at regular intervals (Fig. 2A,C). Bouin's fixative is especially suited to reveal the nuclear and cytoplasmic morphology of the different B lymphocyte compartments in HE-stained paraffin sections. Thus, with this fixation, primary follicles contain a uniform population of small lymphocytes with dark nuclei. However, the colour of the B cell nuclei is somewhat lighter than that of T lymphocytes in the PALS. In addition, the arrangement of the B lymphocytes is not as dense as that observed with T cells in the PALS. Thus, although some experience is needed, the location of primary follicles can be assessed unequivocally by conventional histology (Fig. 2A). In SPF animals secondary follicles are very rare. They are, however, present if the animals are kept under "conventional" conditions or if deliberate immunisation is performed. Inside primary follicles, B cells are distributed within a meshwork of follicular dendritic cells (FDCs).

Whether FDCs are the only resident cell type supporting primary follicles has not been clarified up to now. In any case, FDCs play a decisive role for the secondary stimulation of B lymphocytes and for the formation of germinal centres by retaining antigen-antibody complexes on their surface. Besides FDCs, macrophages are a normal constituent of primary follicles. However, follicular macrophages differ from red pulp macrophages, for example, by their negligible staining with MAb ED1 against macrosialin (CD68) (Fig. 4A).

B lymphocytes in primary follicles may be detected by their expression of the high molecular weight form of the leukocyte common antigen (LCA, CD45R) as visualised by MAb Ox33, by the simultaneous expression of surface IgM and IgD and by several other antigens (Fig. 2C). Primary follicles are regions constantly passed by B lymphocytes, i.e. zones of B lymphocyte migration. Secondary follicles with germinal centres may arise when an immune reaction has led to the formation of antibodies and the deposition of antigen-antibody complexes on the surface of FDCs. The formation of secondary follicles implies that a few recirculating B cells primed at extrafollicular sites meet their cognate antigen in the immune complexes retained on FDCs. In this case the antigen-specific B cells are arrested from further migration. The cells enlarge, proliferate and the affinity of their surface immunoglobulin (Ig) receptors is further increased by hypermutation of the genes encoding the Ig hypervariable antigen-binding regions.

Germinal centre B cells which fail to produce functional Ig receptors by this mechanism or which experience inappropriate ligation of surface Ig cannot be rescued and eventually die by apoptosis. These apoptotic cells are removed by macrophages, called the "tingible body macrophages" of germinal centres, because of the dark-staining phagocytosed nuclear remnants in their cytoplasm.

The proliferating and differentiating B lymphocytes of the germinal centre have larger nuclei and more cytoplasm than the surrounding B cells. In contrast to human splenic secondary follicles, it is often very difficult to recognise a pale germinal centre within HE-stained rat follicles. In rats, B lymphocytes in the germinal centre are only slightly larger and paler than the B cells outside. The "light zone" of the germinal centre (see Sect. 5.1.3) is especially difficult to visualise in rats. This zone often contains a mixture of germinal centre cells and small dense B lymphocytes which are indistinguishable from the B cells of the corona (Fig. 2B).

Those B cells which do not meet cognate antigen are not arrested within the respective follicle and leave again. They form the dark-staining corona or mantle zone as they pass around the germinal centre (Fig. 2B). In other words, the corona represents the equivalent of the primary follicle, pushed to the periphery by the developing germinal centre (for further details see Sect. 5.1.3). During primary immune reactions, such as acute allograft rejection in rats, the follicles keep their hemispherical or crescent shape when germinal centres arise. This is due to the fact that the corona surrounds the germinal centre on the outside towards the marginal zone and red pulp, but is absent in the space between the PALS and the germinal centre (Figs. 1, 2D). In rats and man, the B cells of the corona are characterised by the simultaneous expression of surface IgM and IgD (Gray et al. 1982). In addition, in rats they carry an especially high amount of the CD45R (Ox33) surface molecule (Fig. 2D). These strongly Ox33-positive cells are also present as a thin row in the oPALS (Fig. 2C, D).

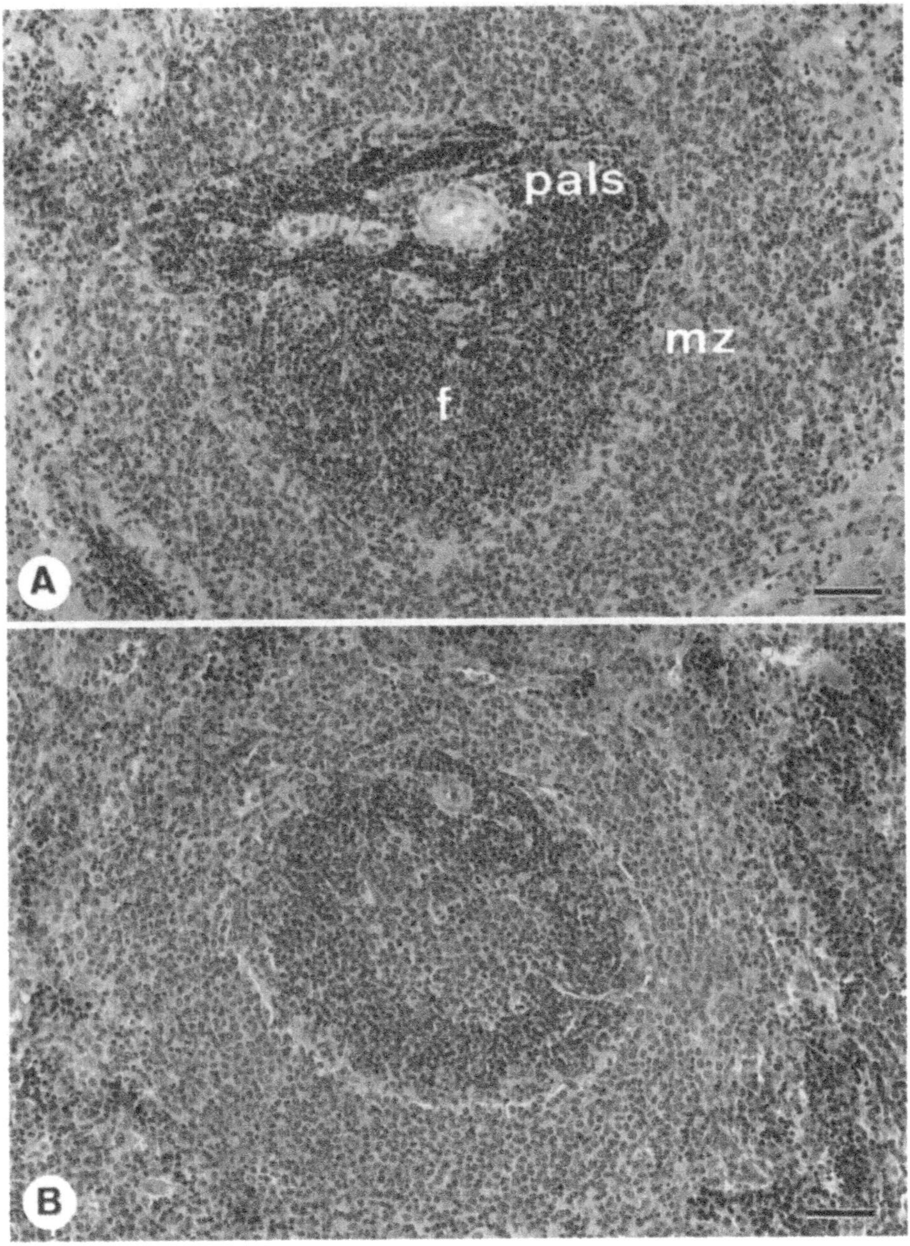

Fig. 2. Overview of the splenic white pulp and distribution of B cells in three-month-old LEW rats kept under specified pathogen-free (A,C) or under conventional (B,D) conditions. **A** HE-stained paraffin section showing periarteriolar lymphatic sheath (*pals*), primary follicle (*f*) and marginal zone (*mz*). The marginal sinus and the marginal metallophilic macrophages form a pale band delimiting the follicle from the marginal zone. **B** HE-stained paraffin section of a secondary follicle showing a light germinal centre surrounded by the dark corona. The PALS is not clearly distinguishable. The marginal sinus separating the corona and the PALS from the pale marginal zone is distended due to a sectioning artefact.

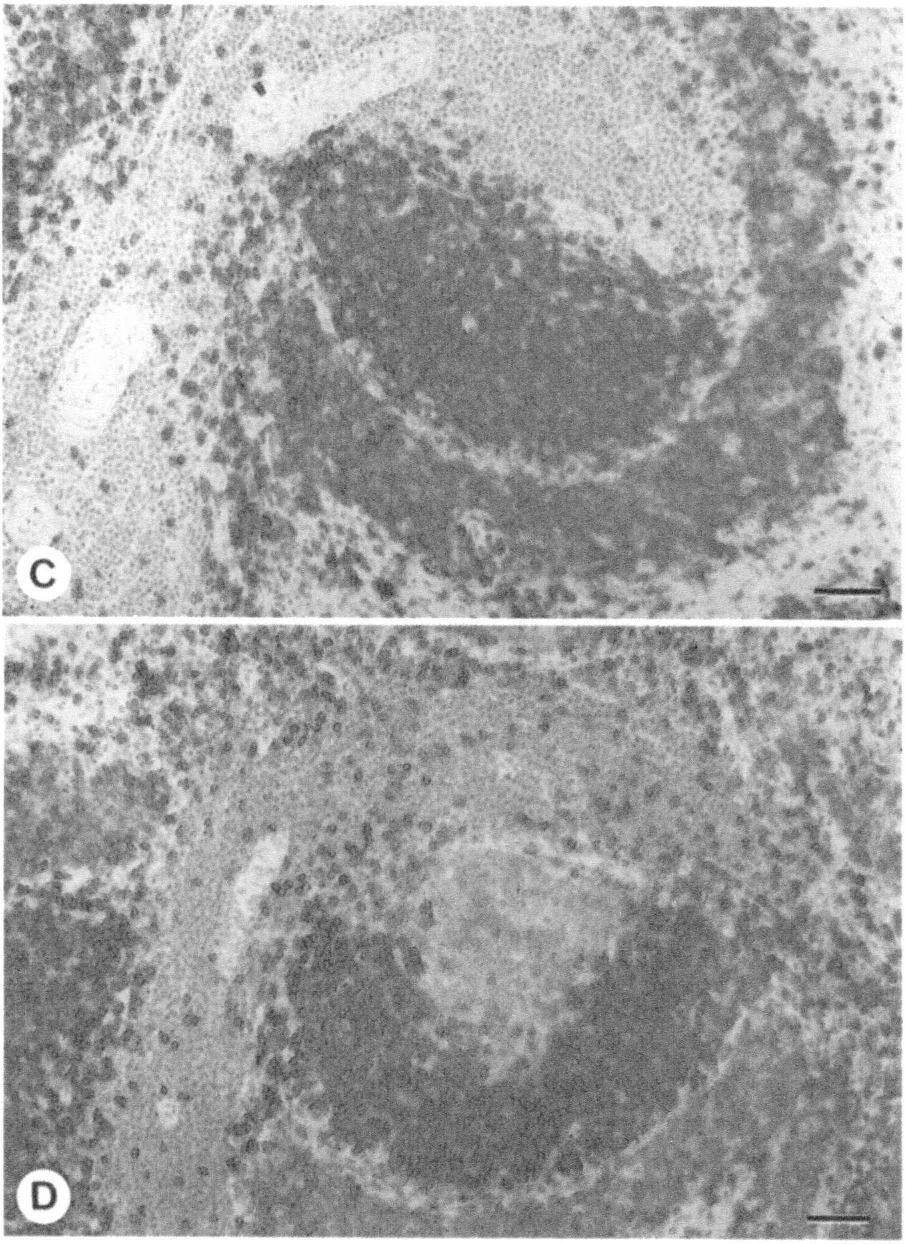

Fig. 2 *(continued)*. **C** Cryostat section stained with MAb Ox33 to demonstrate B cells. Ox33 strongly reacts with B cells of the primary follicle and somewhat less intensely with marginal zone B cells. The majority of lymphocytes in the PALS do not react, but single migratory B cells are present in the outer PALS. This region is cut tangentially in the *left* part of the figure. **D** Cryostat section stained with MAb Ox33 to demonstrate B cells. In a secondary follicle Ox33 strongly reacts with B cells of the corona and somewhat less intensely with marginal zone B cells. B cells of the germinal centre are only faintly stained. The outer PALS in the *upper* part of the figure is cut tangentially, so that the B cells are revealed. ABC technique using DAB chromogen (C,D). Bar = 50 μm (A-D).

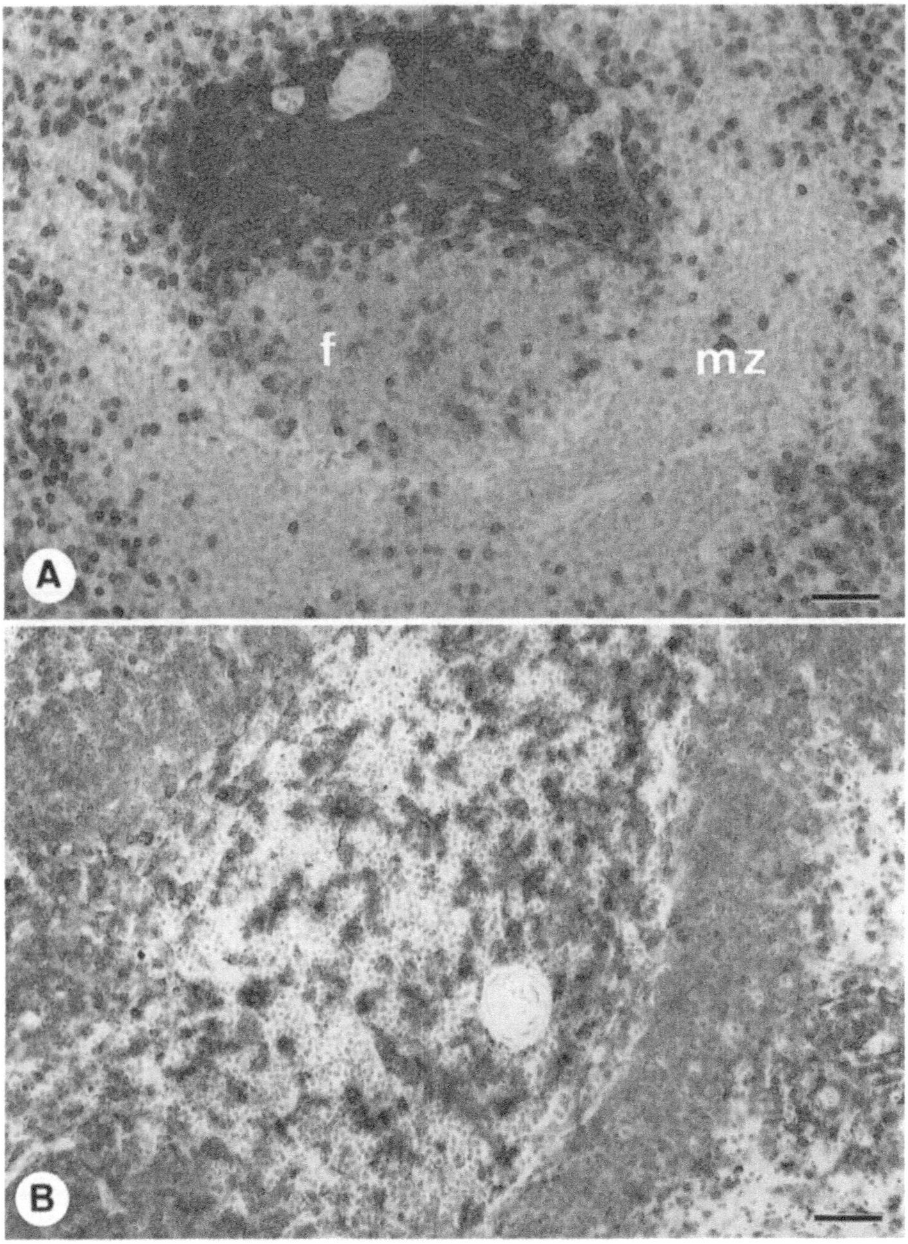

Fig. 3. T cells, interdigitating dendritic cells and myofibroblasts in the rat splenic white pulp. **A** Cryostat section of PALS and a primary follicle stained with MAb R73 to reveal cells bearing the alpha/beta T cell receptor. The PALS is strongly reactive, but there are also a number of T cells in the primary follicle and T cells migrating across the marginal zone. A large number of T cells lodge in the red pulp at the periphery of the figure. Indirect immunoperoxidase technique using DAB chromogen. **B** Cryostat section of PALS and marginal zone stained with MAb Ox3 reacting with MHC class II molecules. The B cells of the marginal zone and interdigitating dendritic cells in the PALS are positive. Indirect immunoperoxidase technique, using DAB chromogen.

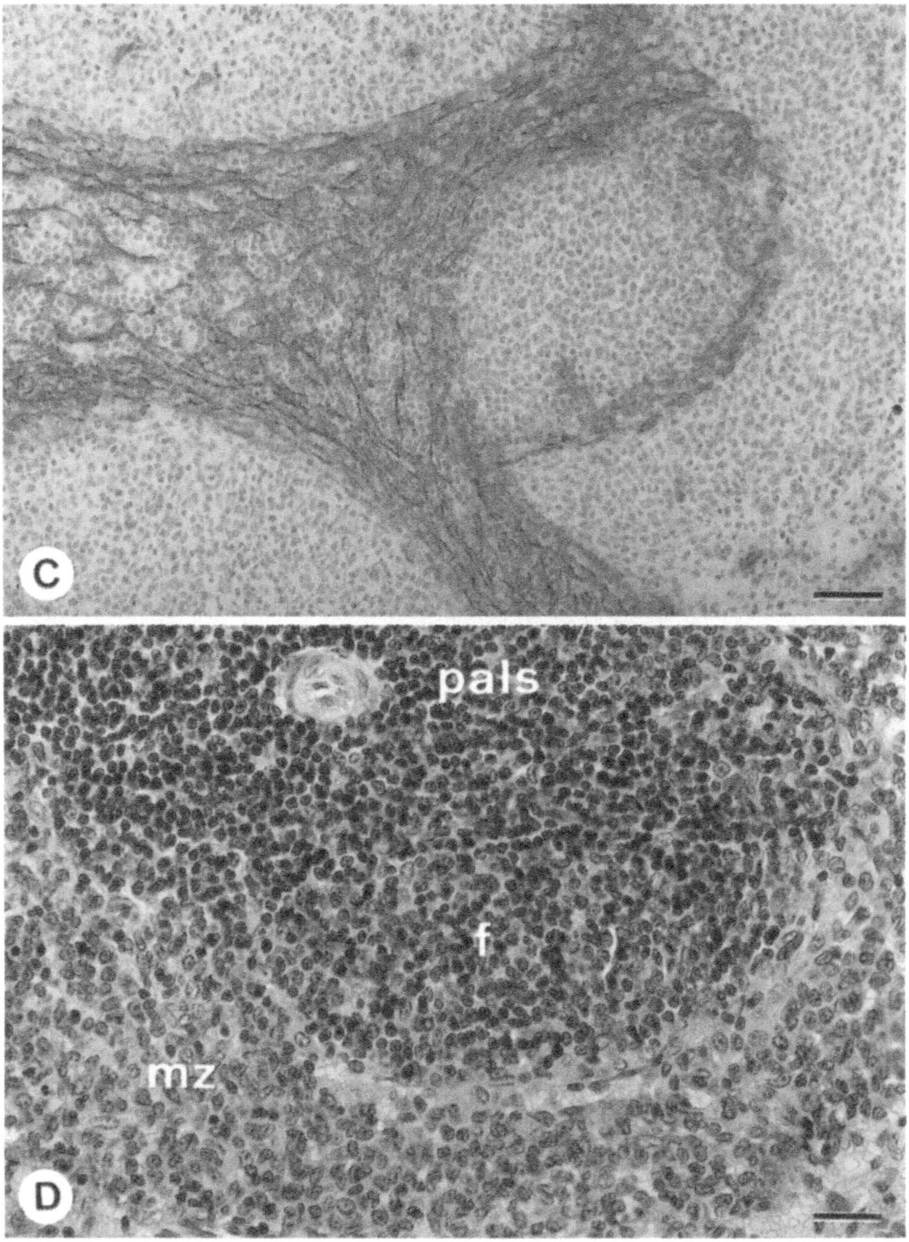

Fig. 3 *(continued)*. **C** Cryostat section of PALS and primary follicle stained with MAb asm-1 directed against smooth muscle alpha-actin. Myofibroblasts in the PALS and in the vicinity of the marginal sinus are positive, the follicle and the marginal zone remain unstained. Note the denser meshes of myofibroblasts in the outer PALS. ABC technique using DAB chromogen. **D** HE-stained paraffin section of PALS, primary follicle (*f*) and marginal zone (*mz*) demonstrating the marginal sinus and its endothelial cells between the follicle and the marginal zone. All specimens were taken from 3-month-old male LEW rats kept under specified pathogen-free conditions. Scale bars (A-C), 50 µm; (D), 25 µm.

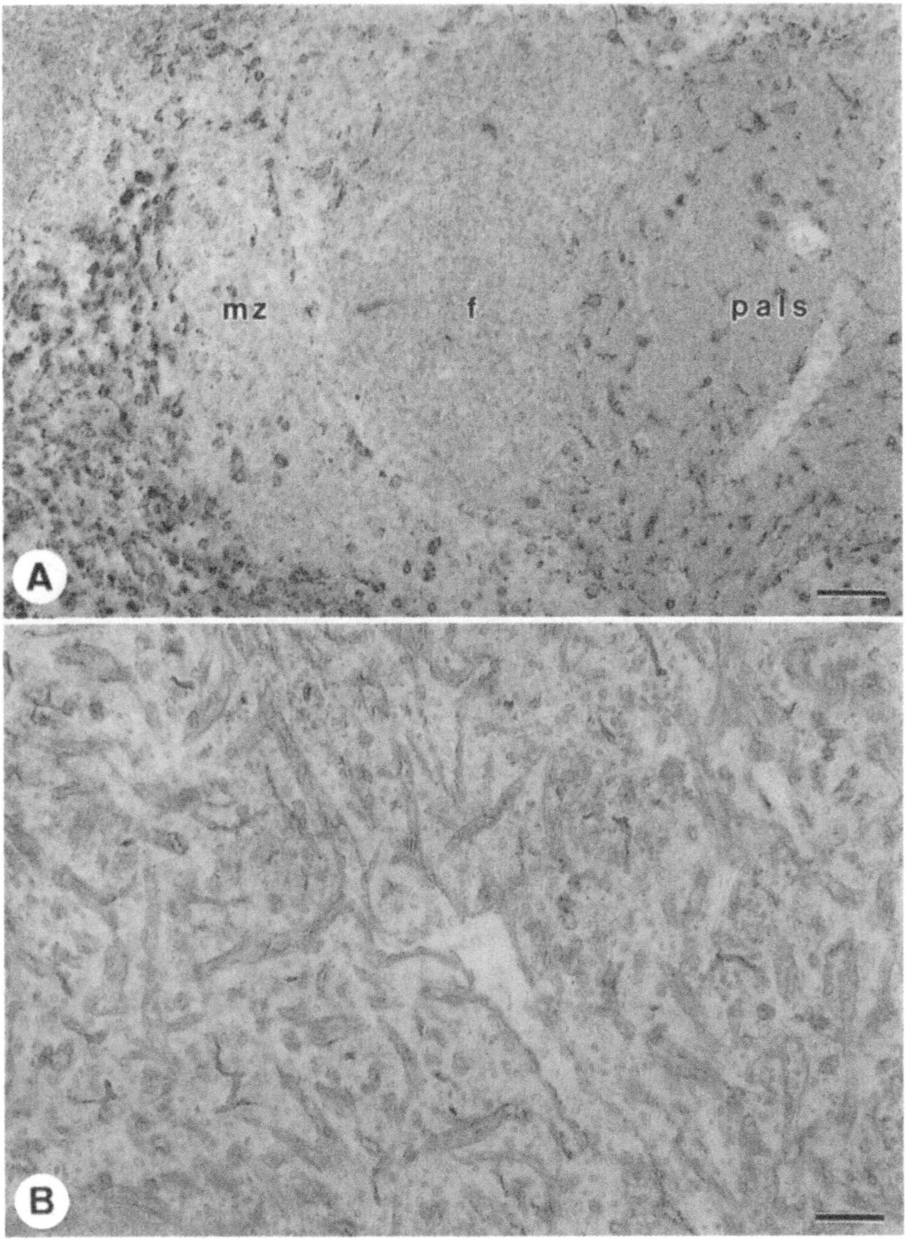

Fig. 4. Distribution of macrophages in the splenic white pulp and visualisation of red pulp vessels in rats. **A** MAb ED1 directed against a lysosomal antigen similar to macrosialin and CD68 stains a primarily intracellular antigen in macrophages and interdigitating dendritic cells in the PALS and weakly reacts with a few macrophages in the primary follicle (*f*) and in the marginal zone (*mz*). A minority of marginal metallophilic macrophages may be reactive. Red pulp macrophages are strongly positive. Indirect immunoperoxidase technique using DAB chromogen. Bar = 50 μm. **B** MAb HIS52 stains the endothelium of red pulp sinuses and other vessels in the rat spleen. ABC technique using DAB chromogen. Bar = 50 μm.

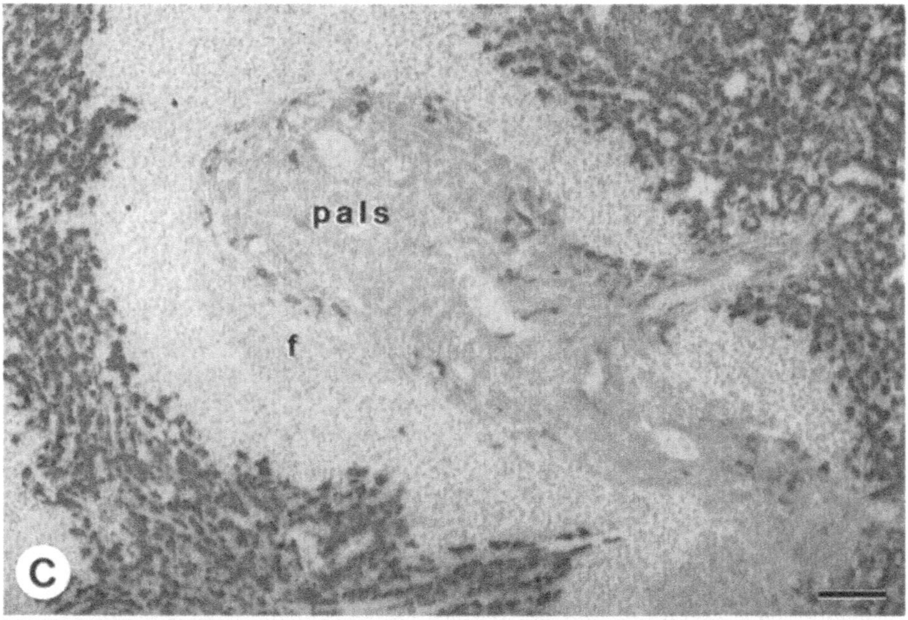

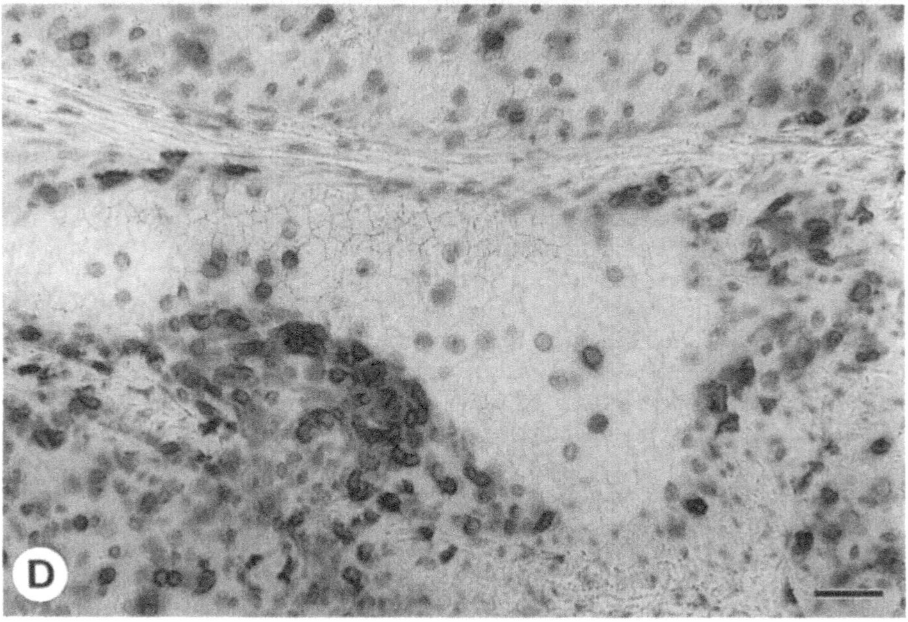

Fig. 4 *(continued)*. **C** MAb ED2 reacts with a surface antigen on red pulp macrophages and on a small population of macrophages located in the outer PALS. These macrophages accompany a marginal zone bridging channel leaving the PALS for the red pulp in the *right* part of the figure. Marginal metallophilic macrophages are negative as revealed by the unstained periphery of the primary follicle (*f*). Bar = 100 μm. **D** T cells expressing the alpha/beta T cell receptor (MAb R73) in the process of migrating across the smooth muscle layer of a major vein in the splenic red pulp. Indirect immunoperoxidase technique using DAB chromogen (C,D). Bar = 25 μm. Cryostat sections (A-D) were taken from 3-month-old LEW rats kept under specified pathogen-free conditions.

3.1.3 Marginal Zone

The MZ is the third region distinguishable in the splenic white pulp of rats. It surrounds both PALS and follicles and forms a broad and well-demarcated border between white and red pulp (Figs. 1, 2A–D). In HE-stained sections the MZ stands out as a rather clear region due to the large B lymphocytes with ample cytoplasm and light nuclei, which form the predominant cell population (Fig. 2A, B). In rats and mice, the majority of MZ B lymphocytes represent a special type of B memory cell, characterised by surface IgM expression but absent or low expression of IgD (Liu et al. 1988; Gray et al. 1982; Kumararatne et al. 1981). It is assumed that the MZ memory B cells represent the decisive cell population responsible for immediate antibody responses against polysaccharide constituents of bacterial capsules (see Sect. 5.2). Thus, the MZ may form the sepsis-protective compartment of the spleen (MacLennan and Liu 1991). It has been shown that MZ B cells are a sessile cell population slowly replenished from the recirculating pool of B cells. However, after cross-linking of surface Ig or during immunological emergencies, MZ B cells quickly leave their compartment for further differentiation (see Sects. 5.1.4 and 5.2).

The MZ is a multipurpose compartment which also serves the immigration of recirculating lymphocytes into the white pulp and the entry of antigenic material from the blood. To fulfil this function, the MZ belongs to the open part of the splenic blood circulation. It lacks endothelia and allows erythrocytes and leukocytes access to all resident cells (Altschul and Hummason 1947; Snook 1964; Veerman and van Ewijk 1975). Recirculating B and T lymphocytes initially arrive in the MZ when entering the white pulp (Nieuwenhuis and Ford 1976). From the MZ, T and B cells move into the oPALS to continue their divergent migration (for further details see Chap. 6). The border between the PALS and the follicles on one side and the MZ on the other side is sharply demarcated in rats by the marginal sinus, a rather irregular capillary vessel (Fig. 1, 3D)(Snook 1950). Larger particles, such as dyes, injected intravenously, quickly extravasate through the wall of the marginal sinus into the MZ, but not into the PALS or the follicles (Sasou et al. 1976).[1] Thus, there is an asymmetric permeability of the marginal sinus walls due to a discontinuous endothelium which occurs on the MZ side of the sinus. It has been assumed that some proportion of the erythrocytes, thrombocytes and white blood cells passing through the marginal sinus can more or less freely access the MZ via this endothelium (Veerman and van Ewijk 1975). Whether there are additional vessels ending in the MZ is not clear. The marginal sinus represents a meshwork of spaces lined by endothelia, which does not form a barrier between the compartments. Thus, lymphocytes may move from the MZ to the PALS or to the follicles between the meshes of the marginal sinus (Sasou 1976). A layer of myofibroblasts expressing smooth muscle alpha-actin accompanies the marginal sinus. Immunohistology reveals these myofibroblasts in the periphery of follicles as a band-like structure, which continues into the oPALS (Fig. 3C).

[1] Nossal et al. (1966) have, however, shown that labelled protein antigens slowly "migrate" from the MZ into the interior of follicles within hours after i.v. injection. How this migration is accomplished has not been clarified up to now.

In mice and rats, two specialised types of macrophages reside in the MZ, the "marginal metallophilic macrophages" (MMMϕ) near the marginal sinus and the "marginal zone macrophages" (MZMϕ) scattered throughout the MZ. MMMϕ were first visualised by silver impregnation in rats (Snook 1964). These cells line the follicular and the PALS side of the endothelial wall of the marginal sinus (Fig. 1). In mice, MMMϕ can be specifically stained with MAbs (Kraal 1992), however, a MAb specific for rat MMMϕ is not yet available. Thus, in rats MMMϕ and MZMϕ are indistinguishable up to now. Both cell types are characterised immunohistologically by their strong cell-surface expression of sialoadhesin as visualised by MAb ED3 and by staining with MAb KiM9R (Fig. 5A) directed against a thus far unknown macrophage antigen. Sialoadhesin is a rather unusual member of the immunoglobulin protein superfamily with an extracellular portion consisting of 17 immunoglobulin domains (Crocker et al. 1994). As its name indicates, sialoadhesin binds to sialic acid-containing glycoconjugates. In rats, it is supposed to act as a cell adhesion molecule mediating interactions among recirculating lymphocytes, MMMϕ and MZMϕ (van den Berg et al. 1992). The expression of sialoadhesin is restricted to macrophages and strong expression is normally only present in a very limited number of specialised cells (Dijkstra et al. 1985a). In addition to sialoadhesin, rat MMMϕ and MZMϕ both express acid phosphatase activity (Fig. 5C). The CD68 lysosomal membrane protein, or macrosialin detected by MAb ED1, which is a rather specific immunohistological marker of macrophages in rats and humans, is, however, hardly expressed in both macrophage populations in rats (Fig. 4A). It is clear that the MZ contains a population of reticular fibroblasts, but detailed information about these cells is lacking.

3.2 Red Pulp and Splenic Vessels

The red pulp of rats is composed of sinuses and splenic cords (Fig. 4B). The border of the MZ towards the red pulp is easily determined although it is not formed by any special structure (Fig. 2A). The splenic sinuses are regarded as representing the "closed" part of the splenic circulation. It is assumed that the walls of the splenic sinuses in rats are similar to those in humans (Chen and Weiss 1973). Thus they consist of endothelia with intercellular slits, which are fixed to "hoops" or ring-fibres of basement membrane-like material traversing the endothelia on their basal surface. The splenic cords surrounding the sinuses contain a large number of mature macrophages (Fig. 4A,C) together with B and T lymphocytes, erythrocytes, thrombocytes, granulocytes, some plasma cells and haematopoietic cells within a meshwork of fibroblasts. The cords form the second part of the "open" circulation pathway in the rat spleen in addition to the MZ. The splenic cords are supplied by small arterioles which represent end or side branches of the central arterioles (Satodate et al. 1986). These vessels discharge their blood into the cord spaces lined by fibroblasts. The blood returns from the cords to the vasculature via the slits located between the sinus endothelial cells, i.e. by entering the sinuses from the outside (MacDonald et al. 1987). It is unknown whether there are other means of drainage from the cords.

In rats, the current model of splenic vasculature is the following: several branches of the splenic artery enter the organ parenchyma and give rise to central arterioles.

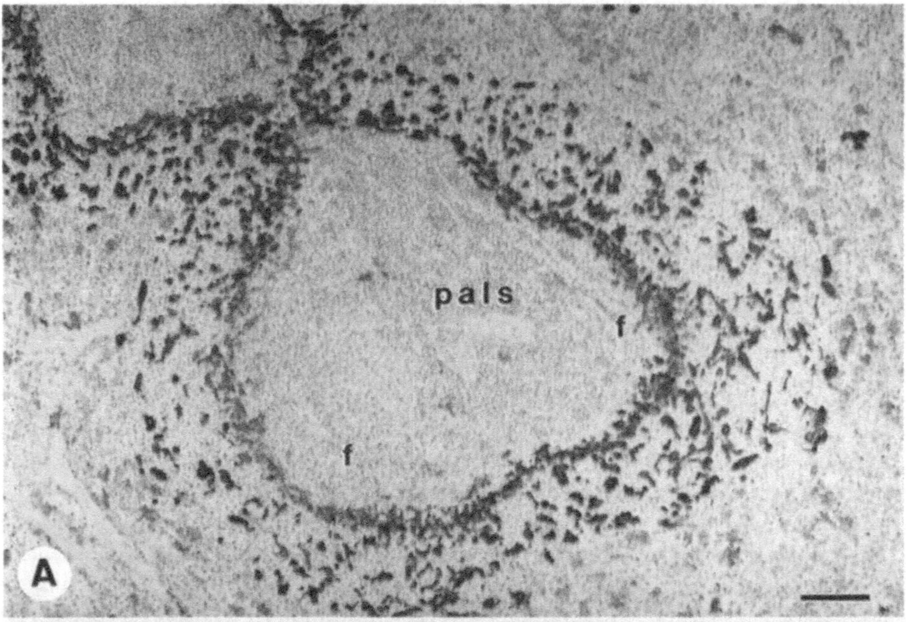

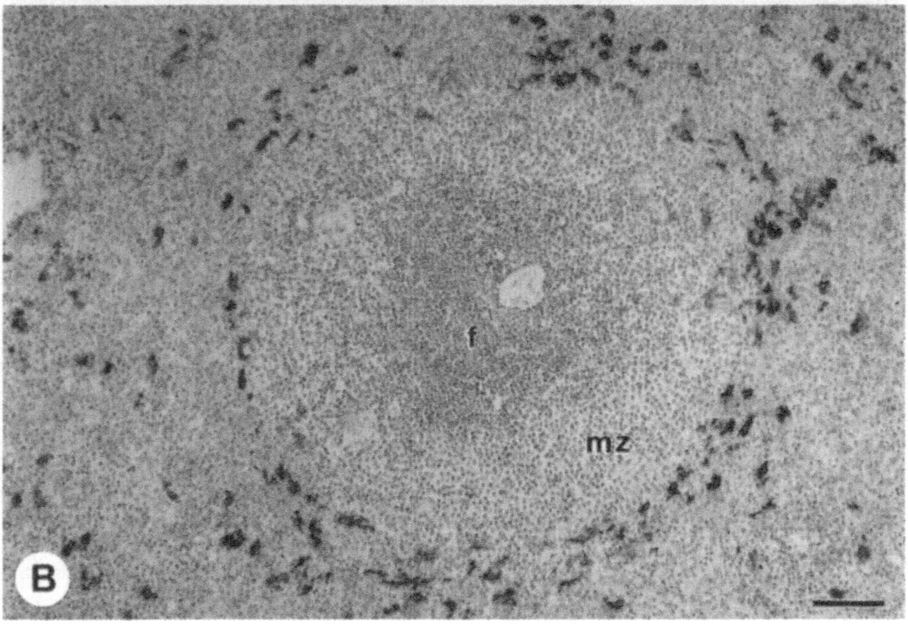

Fig. 5. Comparison of macrophages in the marginal zone of rat and human splenic white pulp. **A** Cryostat section of rat spleen stained with MAb KiM9R. Similar to ED3, KiM9R reacts with marginal metallophilic macrophages and marginal zone macrophages surrounding the PALS (*pals*) and primary follicles (*f*). Indirect immunoperoxidase technique using DAB chromogen. Bar = 100µm. **B** Paraffin section of a human spleen stained with MAb HSN1 against human sialoadhesin. In this unusual specimen strongly HSN1-positive macrophages are situated in the perifollicular zone outside the marginal zone (*mz*). There are no sialoadhesin-positive cells in the follicle and in the marginal zone and no marginal metallophilic macrophages. *f*, primary follicle. ABC technique using DAB chromogen. Bar = 80 µm.

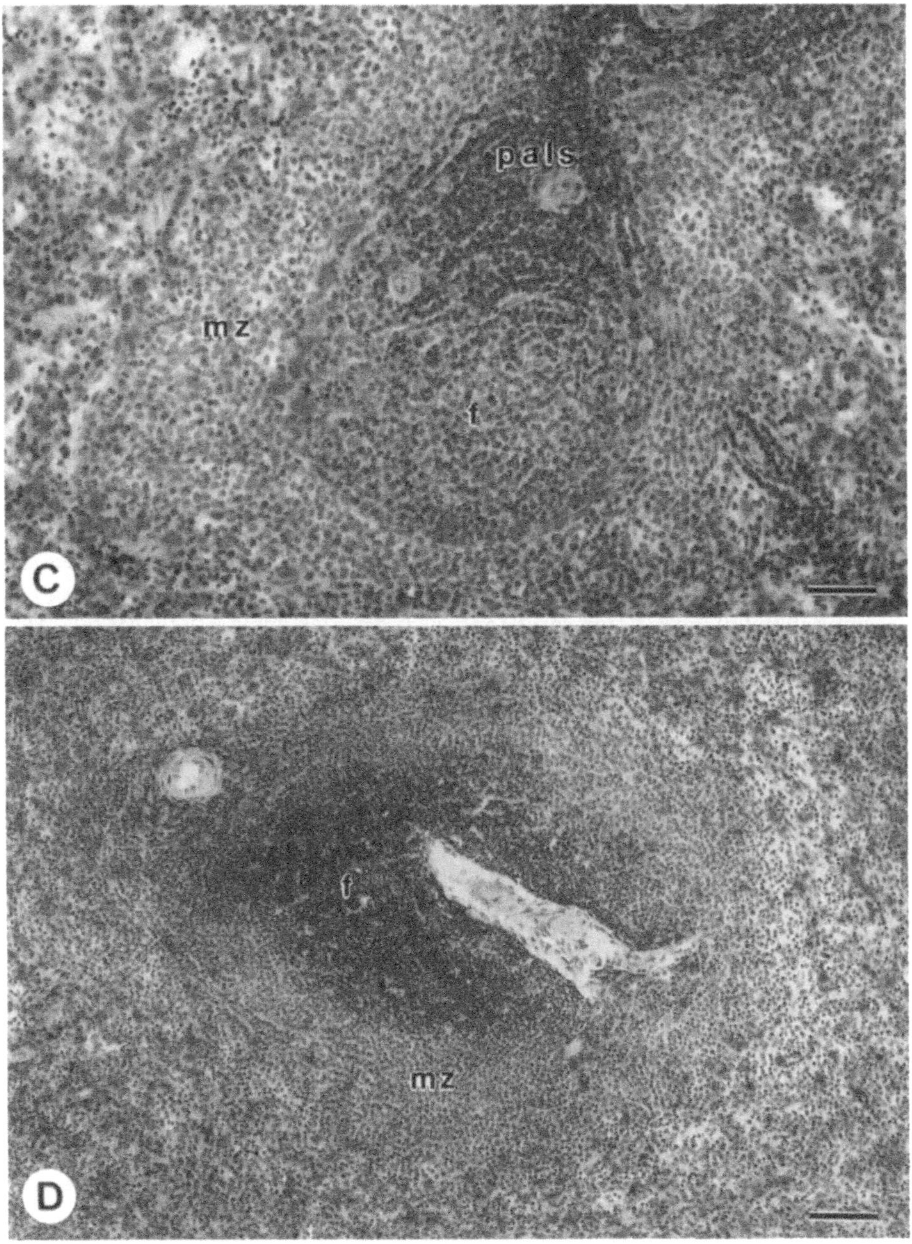

Fig. 5 *(continued)*. **C** Acid phosphatase activity in cryostat section of rat PALS (*pals*), primary follicle (*f*) and marginal zone (*mz*). Enzyme activity is present in marginal metallophilic macrophages and, somewhat more weakly, in marginal zone macrophages. Some macrophages in the PALS and in the follicle are also stained. Bar = 50 μm. **D** Cryostat section of human spleen (same specimen as in B) stained for acid phosphatase activity. Although there are acid phosphatase-positive macrophages in the primary follicle (*f*) and in the marginal zone (*mz*), marginal metallophilic macrophages are not revealed. Bar = 100 μm. 3-month-old male LEW rat kept under SPF conditions (A,C). 59-year-old male patient with pancreatitis (B,D)

Trabecular arteries, as observed in man, are either extremely short or do not exist at all. In fact, there are no trabeculae in the rat splenic red pulp, but only small strands of smooth muscle cells. The central arterioles release capillaries at right angles which either traverse the follicles or the PALS to end up in the marginal sinus. The capillaries often branch within the follicles (Fig. 1). In addition, larger arterioles also leave the central arterioles in a similar direction, but traverse the MZ and enter the red pulp. These vessels are designated as "marginal zone bridging channels" or sometimes as "terminal arterioles". They are accompanied by substantial numbers of B lymphocytes and ED2-positive macrophages (Fig. 4C) and, sometimes, by T lymphocytes. The surroundings of the marginal zone bridging channels have been assumed to serve as migration pathways for B lymphocytes entering the red pulp for further differentiation into plasma cells (Mitchell 1973; Kotani et al. 1986; van Rooijen et al. 1989; Van den Eertwegh et al. 1992a; Liu and Banchereau 1996; Liu 1997). How and where the main stem of the central arterioles ends up in the red pulp is not known. In rats "penicillar arterioles" or "sheathed capillaries" do not occur.

While the distribution of T cells in the red pulp is rather even, B cells tend to be clustered around small or "terminal" arterioles. It is not clear whether these vessels are derived from MZ bridging channels or from the terminal portions of central arterioles. In addition to the B cell clusters around arterioles, T and B lymphocytes are scattered throughout the red pulp. Thus, they are distributed over a large area and the compartment of the red pulp cords contains substantial numbers of these cells. How B and T cells arrive in the splenic cords and how they leave again is unknown. The larger veins draining the red pulp have longitudinal strands of smooth muscle cells and some connective tissue in their walls. They appear to represent the equivalent of the "trabecular veins" in humans. The veins often exhibit prominent aggregates of B or T lymphocytes between the endothelium and the smooth muscle layer with clearly visible signs of lymphocyte diapedesis across the endothelium (Steiniger et al. unpublished, Fig. 4D). Whether these arrangements are a consequence of lymphocyte emigration or immigration is open to speculation.

3.3 Summary

In rats, the splenic white pulp consists of three compartments, the PALS, the follicles and the MZ. The PALS is a compartment where T lymphocytes predominate. The PALS also contains specialised myofibroblasts and antigen-presenting dendritic cells, the so-called interdigitating dendritic cells. The oPALS represents a recirculation pathway for B lymphocytes. Primary follicles are migration compartments for B lymphocytes, which may develop into secondary follicles during immune reactions. Secondary follicles consist of a germinal centre formed by a dark and a light zone and a surrounding follicle corona or mantle zone. The corona represents the remainder of the primary follicle. Apart from proliferating and differentiating B lymphocytes, a germinal centre is populated by follicular dendritic cells and by tingible body macrophages. PALS and follicles are separated from the red pulp by a broad MZ. The MZ is a B lymphocyte compartment, which is predominantly inhabited by sessile IgM^+ B memory cells. An irregular blood vessel, the marginal sinus, separates the PALS and the follicles from the MZ. The marginal sinus is accompanied by a chain of special macrophages, the

MMMϕ. In addition, the MZ contains a second scattered macrophage population, the MZMϕ. Both macrophage populations strongly express sialoadhesin. In addition to the rather stationary B memory cells, there are also recirculating B and T lymphocytes which move across the MZ.

4 Microanatomical Compartments of the Human Spleen

4.1 White Pulp

In contrast to rats the different compartments of the splenic white pulp in man are much more difficult to recognise in conventional HE-stained sections. In addition, the quantity of red and white pulp tends to be individually variable in humans. In general, the white pulp appears less prominent in human than in rat spleens. However, not only the relationship between red and white pulp, but also the predominance of PALS or follicles in the white pulp differs among individuals. In general, the follicles form the most prominent part of the white pulp in man. The PALS is only sparsely distributed (Fig. 6A). Most spleens investigated in our study did not show full-blown secondary follicles, but only primary follicles or follicles with degenerating germinal centres. This phenomenon is most likely influenced by the age and the immunological state of the individual. In adult humans "remnants" of germinal centres may normally prevail, because fully developed germinal centre reactions do not last long if there is no continuous supply of antigen.

4.1.1 Periarteriolar Lymphatic Sheath

Major accumulations of T lymphocytes are only observed around the larger splenic arteries and arterioles in humans. The smaller arterioles are only surrounded by minor numbers of T cells (Fig. 7E). These are the vessels, which are also associated with follicles. The arrangement of PALS and follicles with respect to the "central arterioles" is different from rats. Thus, in the vicinity of follicles, T cells are often asymmetrically distributed around the arterioles. Sometimes the PALS seems to be totally replaced by the B cells of the follicle, so that a "central arteriole" appears to run through the corona or even through the germinal centre of a secondary follicle (Figs. 6C,D, 7E). There may be three alternative explanations for this phenomenon: (a) the PALS may either end in the follicle or (b) it may continue after the vessel has left the other side of the follicle, or (c) the intrafollicular arteriole may in fact represent a side branch of the central arteriole not associated with T cells. In contrast to rats, more arteriolar branches of higher order are associated with the PALS. Thus, the term "central arteriole" is difficult to define, because the vessels surrounded by the T cell compartment differ in size, comprising large and small arteries, arterioles and tiny vessels with only one to two layers of smooth muscle cell in their walls. A highly variable arrangement of follicular

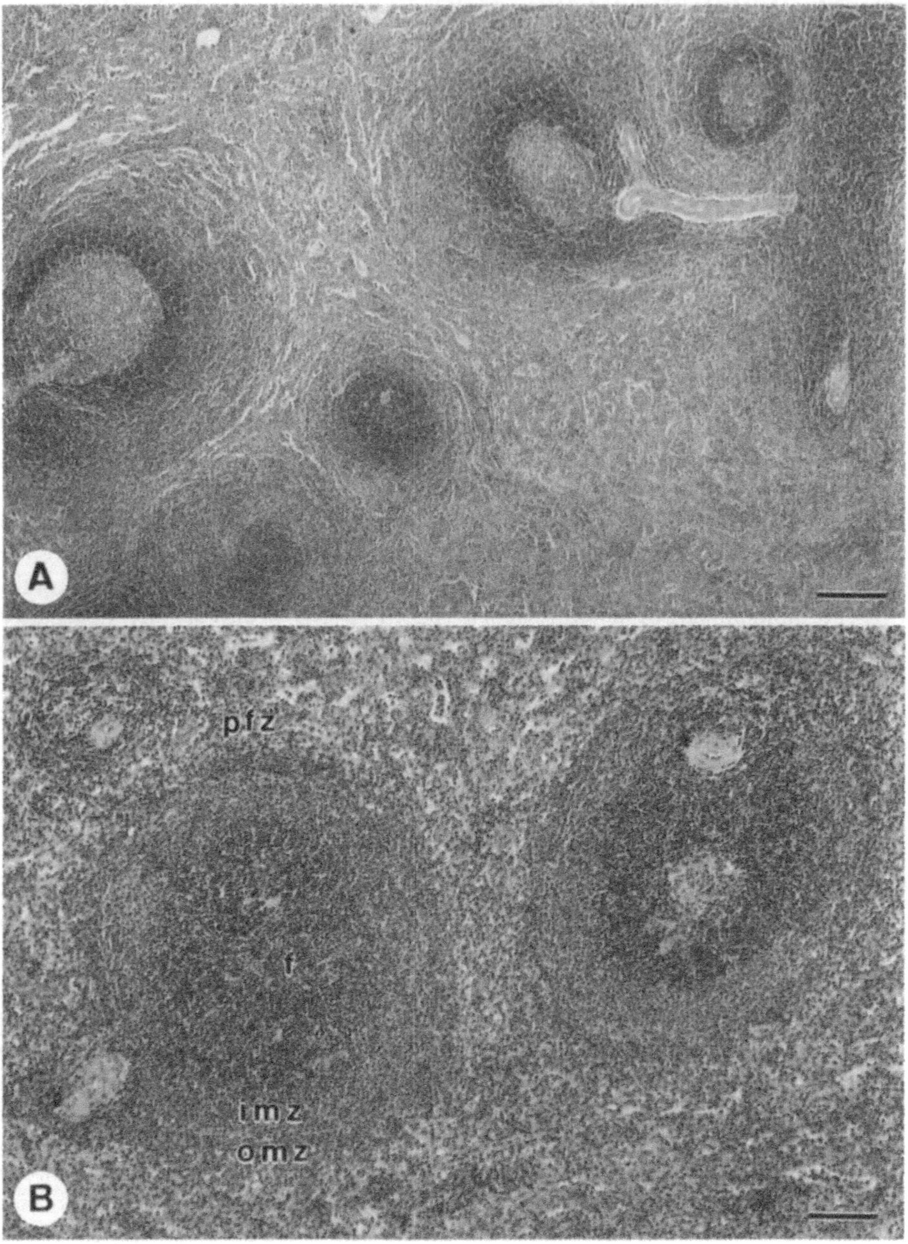

Fig. 6. Distribution of immunoglobulin and macrophages in paraffin sections of the human splenic white pulp. **A** 15-year-old girl with splenic trauma. The full-blown asymmetric secondary follicles contain germinal centres with dark and light zones. The PALS is only sparsely distributed at the *right* margin of the figure. Bar = 200 µm. **B** 29-year-old female with splenic trauma. The right follicle shows the remnants of a germinal centre surrounded by the dark corona. The left follicle (*f*) represents either a primary follicle or a secondary follicle sectioned tangentially. The marginal zone is divided into an inner marginal zone (*imz*) and an outer marginal zone (*omz*) by a row of pale-staining myofibroblasts. The lymphocytes of the omz appear darker than those of the imz. The perifollicular zone (*pfz*) continues into the red pulp. Bar = 100 µm. HE-staining (A,B).

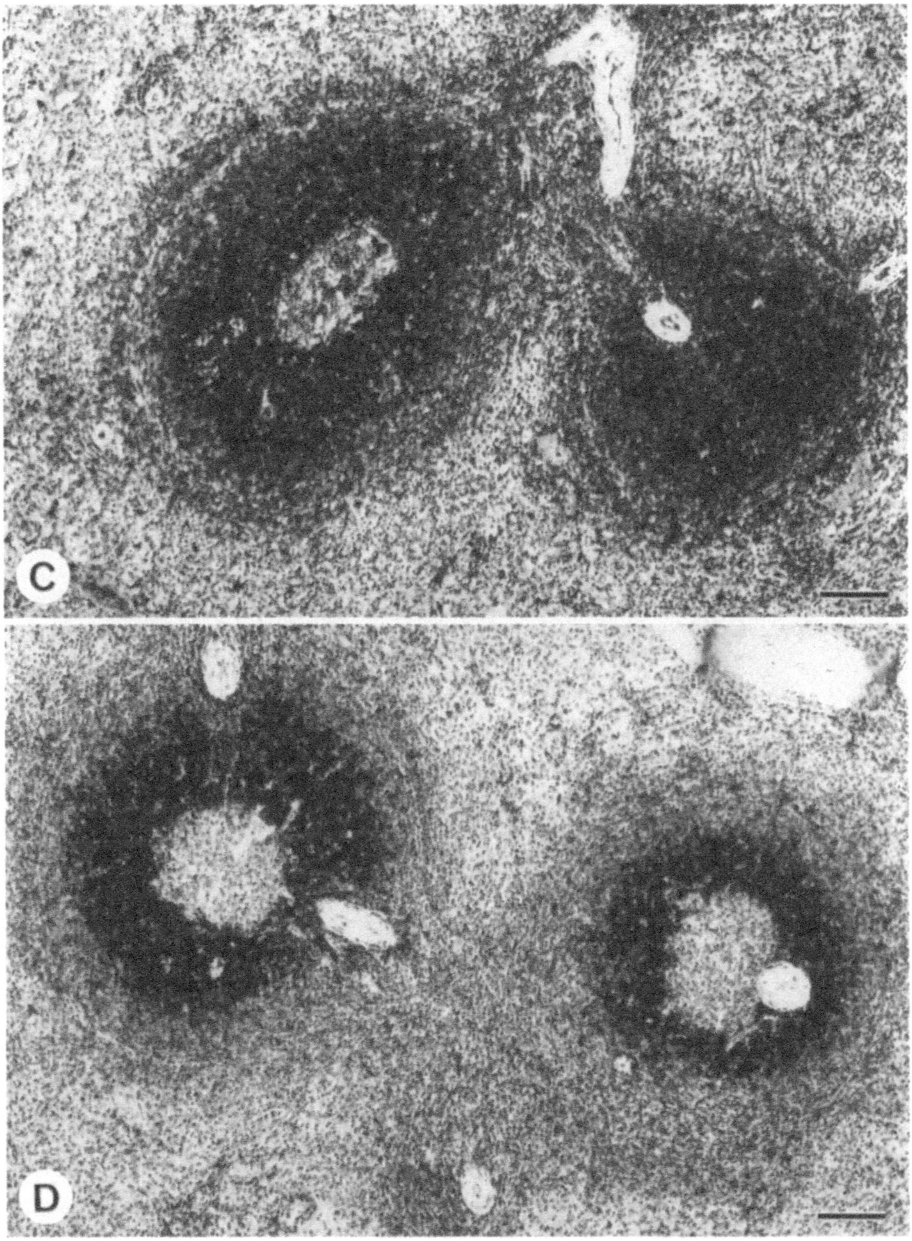

Fig. 6 *(continued).* **C** Two follicles stained dark violet for presence of IgM. The IgM-positive cells of the corona continue into the marginal zone without interruption. The IgM-positive marginal zone is divided into an inner part and an outer part by unstained myofibroblasts. IgM expression is reduced in the germinal centre. Same specimen as in B. **D** Two follicles stained dark violet for IgD. The germinal centres are unstained, the B cells of the corona are strongly positive, while the marginal zone exhibits only very few IgD-positive cells. In the left follicle the pale myofibroblasts are clearly visible. Note that the B cells between the myofibroblasts and the corona are IgD-negative, while in C they stain for IgM. Same specimen as in B. ABC technique using VIP chromogen (C,D). Bars = 100 μm (C,D).

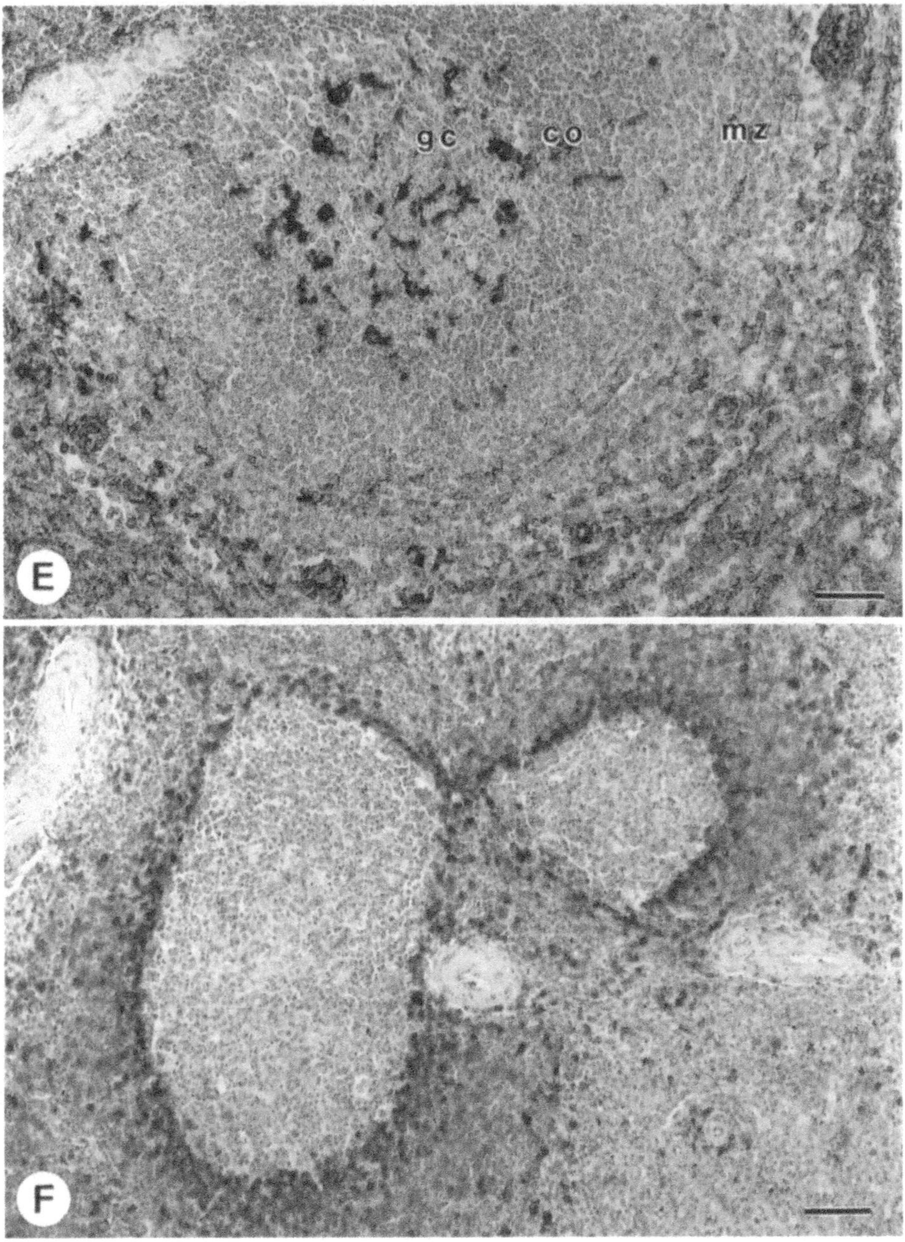

Fig. 6 *(Continued)*. **E** Distribution of KiM1-P (CD68)-positive macrophages. The germinal centre (*gc*) contains strongly positive tingible body macrophages. More lightly stained macrophages are also present in the corona (*co*) and in the marginal zone (*mz*). The perifollicular zone contains strongly positive cells and cannot be distinguished from the red pulp. KiM1-P-positive capillary sheaths are present at the lower margin and upper right corner of the figure. Same specimen as in B. Bar = 50 µm. **F** Two secondary follicles stained for IgD. In this exceptional specimen the B cells of the marginal zone have a low level of IgD expression, which is clearly distinct from the strong expression in the corona. 28-year-old female with splenic trauma. ABC technique using DAB chromogen (E,F). Bar = 80 µm.

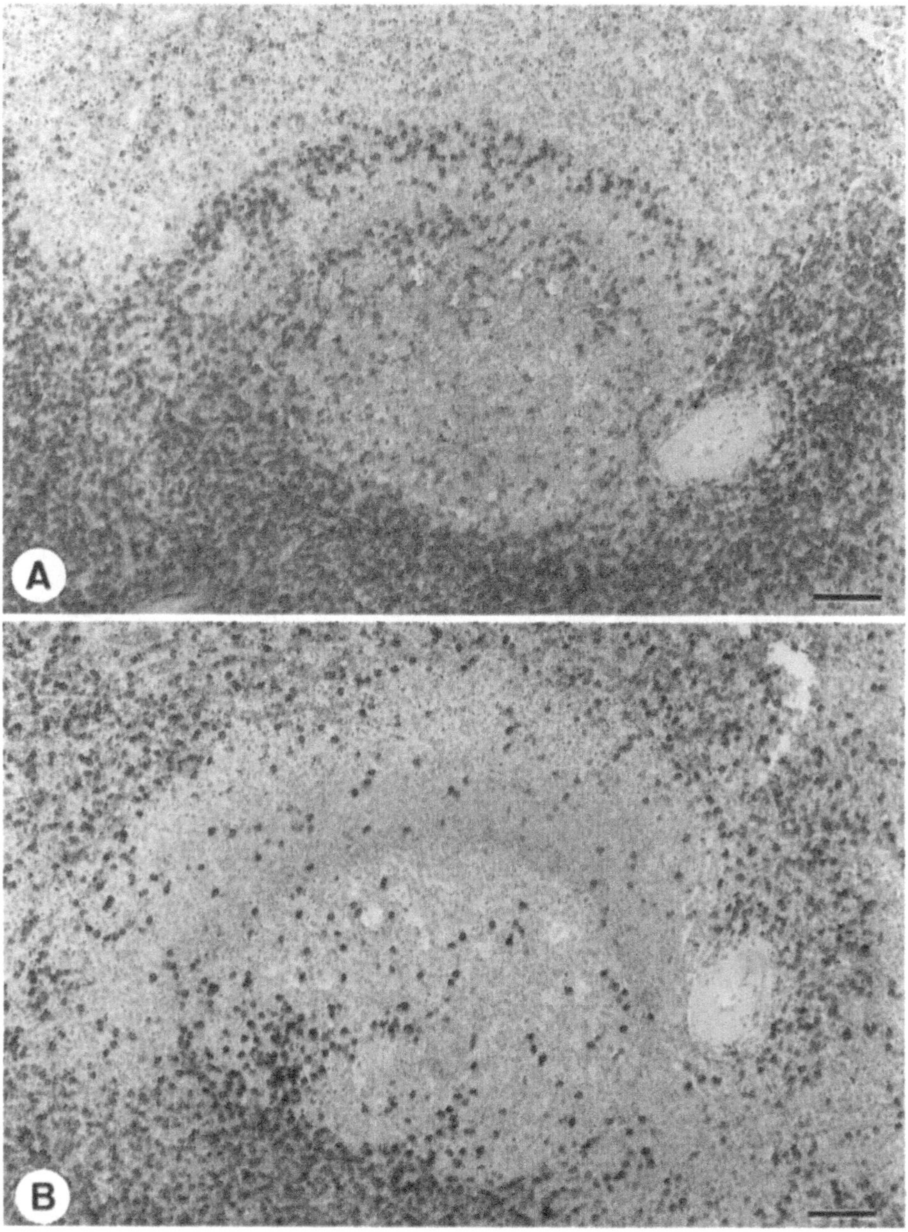

Fig. 7. T cells and myofibroblasts in the human splenic white pulp. **A** CD4$^+$ T cells visualised by MAb 1F6. Dense accumulations of CD4$^+$ T cells are present in the PALS at the lower margin of the figure. The germinal centre contains major numbers of CD4$^+$ T cells bordering the corona. In this specimen (but not always) a row of CD4$^+$ T cells occurs in the marginal zone. Same specimen as in Fig. 6F. **B** Consecutive section of same follicle as in A, stained for CD8 alpha chain expression by MAb 4B11. Very few CD8$^+$ T cells are found in the germinal centre, the corona and the marginal zone. Red pulp sinus endothelia are also stained. ABC technique on autoclaved paraffin sections using DAB chromogen (A,B). Bars = 80 μm (A,B).

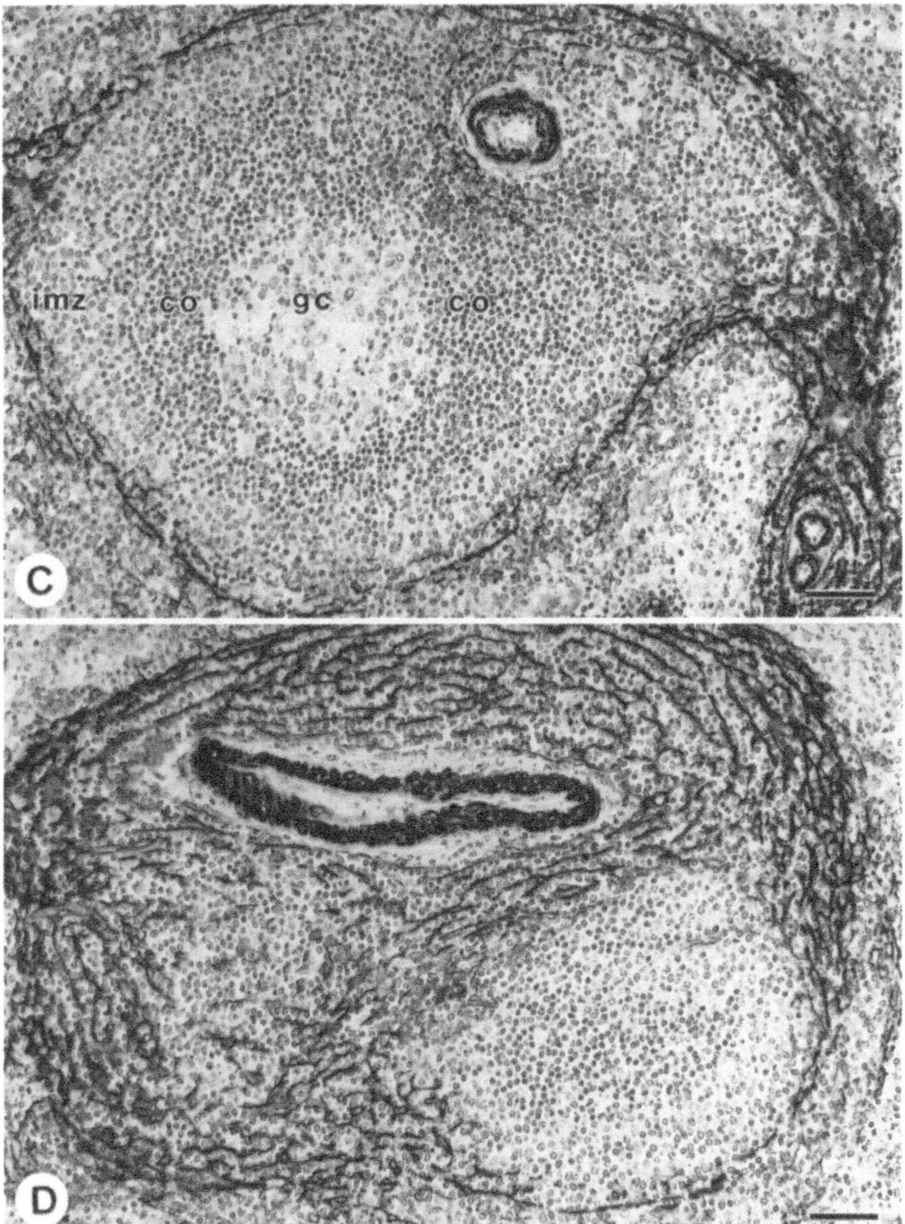

Fig. 7 *(continued)*. **C** Detection of smooth muscle alpha-actin by MAb asm-1. A row of myofibroblasts divides the inner marginal zone (*imz*) from the outer marginal zone. The germinal centre (*gc*) is decaying and exhibits a broad corona (*co*). The myofibroblasts continue into the outer PALS at the *right margin* of the figure. Same specimen as Fig. 6B. **D** PALS and partial section of a follicle stained by MAb asm-1. The myofibroblasts form a concentric network in the PALS of smaller arterioles. If the PALS is larger, there is a reduced or absent staining in the inner PALS. Same specimen as Fig. 6B. ABC technique on paraffin sections (C,D). Bars = 50 µm (C,D).

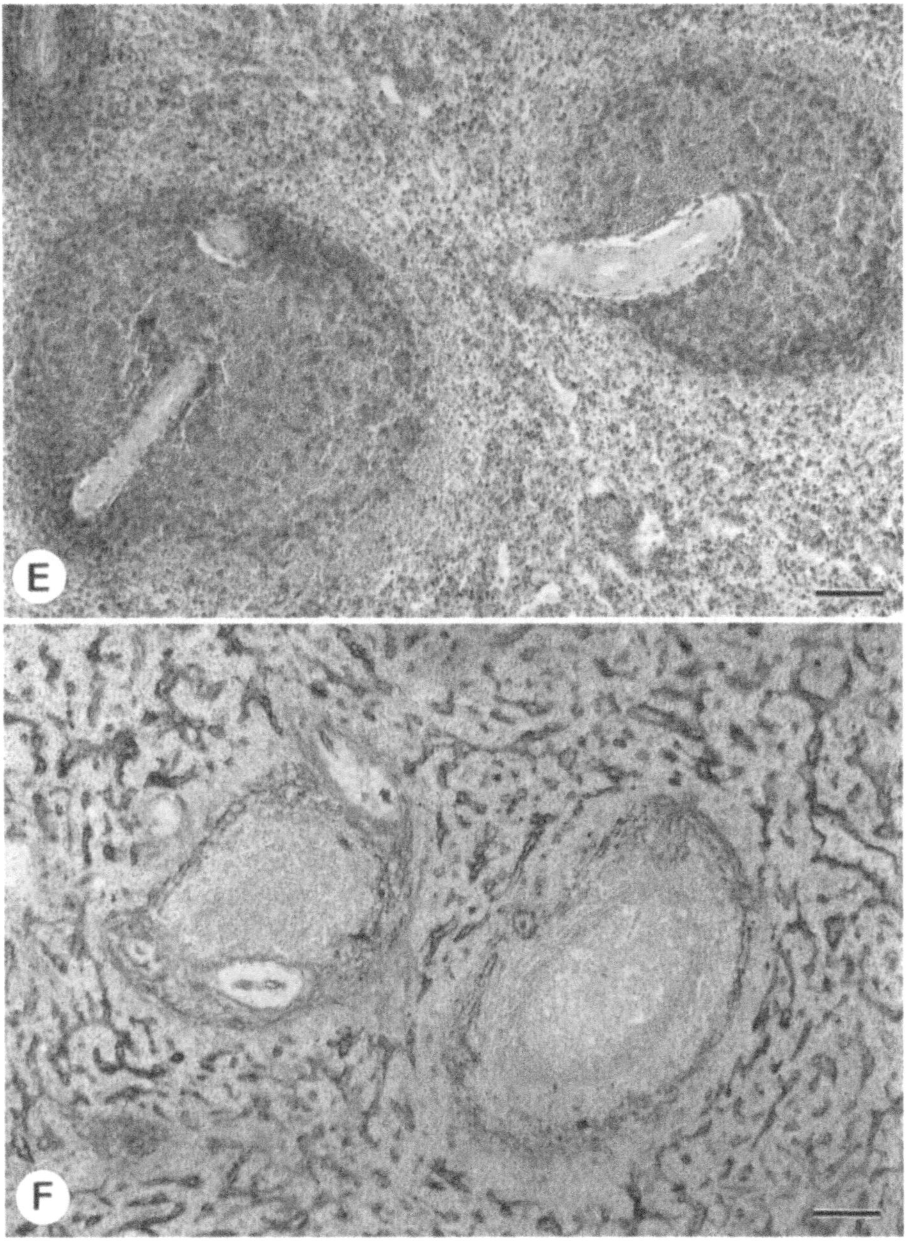

Fig. 7 *(continued).* **E** Demonstration of CD3 by a polyclonal rabbit antiserum. This is a specimen with a very pronounced accumulation of T cells in the marginal zone of two primary (or tangentially sectioned secondary) follicles. In this section the central arterioles appear to lose their PALS when entering the follicle centre. 35-year-old male patient with splenic trauma. **F** Expression of thrombomodulin in two splenic follicles demonstrated by MAb T 1009. The myofibroblasts in the marginal zone and in the PALS are positive. In addition, the endothelium of red pulp sinuses and other vessels is strongly stained. 53-year-old female patient with gastric malignancy. ABC technique on paraffin sections using DAB chromogen (E,F). Bars = 100 μm (E,F).

arterioles in the human spleen was noted early on by Billroth (1861) and the situation has not been clarified since then.

The T lymphocytes in the PALS are positive for CD3 (Fig. 7E) and various other T cell antigens. The majority of the T cells also express the T cell co-receptor CD4 (Fig. 7A), while the alpha chain of the CD8 co-receptor is only present in a minor number of cells (Fig. 7B). The arrangement of MHC class II-positive IDCs is similar to that of rats. The IDCs are also strongly reactive with MAb KiM1-P detecting CD68. Cross-sections of larger areas of the PALS are often recognised by the regular arrangement of the small T lymphocytes which form concentric cell layers. In HE-stained paraffin sections, the layers of lymphocytes are subdivided by pale elongated fibroblast-like cells and fibres. These fibroblasts stain for alkaline phosphatase. In addition, MAb asm-1 directed against smooth muscle alpha-actin reacts with the elongated and branched cells in paraffin sections. Thus, similar to rats, myofibroblasts provide a concentric network in the PALS (Fig. 7C,D). If the PALS is broad, the staining intensity for smooth muscle actin is much higher in the oPALS than in the inner PALS. Sometimes asm-1 reactivity appears to be absent in the inner PALS. The strongly asm-1-positive myofibroblast network of the oPALS continues into the MZ of the follicles as a rather thin structure dividing the MZ into an inner and an outer compartment (see Sect. 4.1.2). This arrangement is different from that found in rats, where the asm-1-positive cells line the marginal sinus. Thus, in humans, primary or secondary follicles distend the periarteriolar network of asm-1-positive cells like balls caught in a net (Fig. 7C,D). Smooth muscle alpha actin-positive fibroblasts are also detected in paraffin sections of human lymph nodes or tonsils. However, in these organs the myofibroblasts exhibit variable staining intensity (see Sect. 8.2). Cryostat sections of human spleens reveal that MAb asm-1 also weakly reacts with branched cells in the red pulp of some, but not all, specimens. The asm-1-strongly positive myofibroblasts in the PALS and the follicular MZ have a rather unusual phenotype. In addition to smooth muscle actin they also stain for thrombomodulin in paraffin sections (Steiniger et al. unpublished, Fig. 7F). The cells are, however, clearly distinct from endothelial cells due to their branched morphology and to the fact that they do not express the CD31 (PECAM-1) molecule, which can be detected on all endothelia in paraffin sections of the human spleen. Even more surprisingly, the same myofibroblasts react with an antibody directed against certain cytokeratins, among them cytokeratin 8, in cryostat sections. Other authors' findings (van den Berg et al. 1993) indicate that the myofibroblasts can also be expected to exhibit an unusual pattern of integrin expression. For a more detailed discussion of these cells see Sect. 8.2.

4.1.2 Follicles

The overall shape of human primary and secondary splenic follicles is globular or ovoid instead of hemispherical and it thus differs from that of rats (Fig. 6A-F). HE-stained sections and paraffin sections stained for the expression of surface Ig and for the CD20 or CD79a B cell-surface molecules show that human splenic secondary follicles are composed of three to four regions (Figs. 6C,D,F, 8, 9A-C).

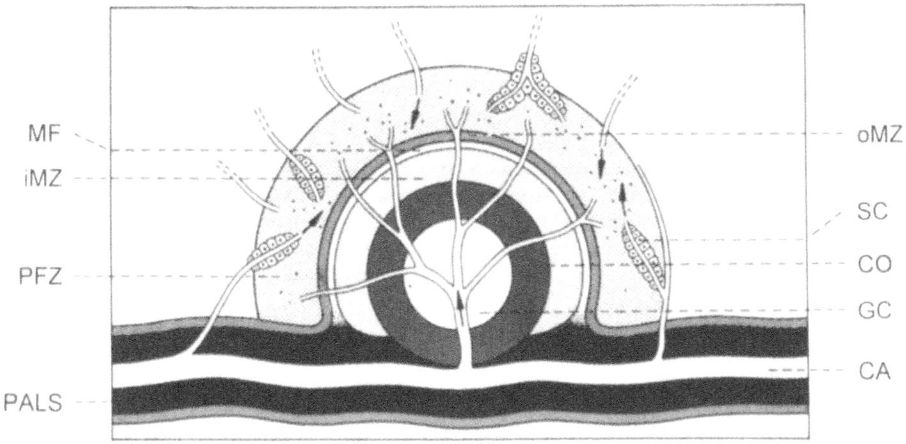

Fig. 8. Schematic drawing of a longitudinal section through the human splenic white pulp. The central arteriole (*CA*) is accompanied by a relatively sparse periarteriolar lymphatic sheath (PALS). Germinal centre (*GC*), corona (*CO*) and marginal zone (*MZ*) represent B cell regions. The marginal zone is subdivided by myofibroblasts (*MF*) into an inner marginal zone (*iMZ*) and an outer marginal zone (*oMZ*). The stippled area at the junction of myofibroblasts and PALS indicates that CD4-positive T cells may accompany the myofibroblasts and that the myofibroblasts continue into the outer PALS. The vascular structures in the perifollicular zone (*PFZ*) are hypothetical: Sheathed capillaries (*SC*) and capillaries from the germinal centre may deliver their blood into open spaces. Distinct borders between corona and marginal zone and between marginal zone and perifollicular zone are absent due to lymphocyte migration. Border lines are only depicted for didactic reasons. Capillaries and zones of the follicle are not drawn to scale. The "central arteriole" may also traverse the corona or the germinal centre.

4.1.2.1 Germinal Centre

Asymmetrical germinal centres with a dark zone formed by centroblasts oriented towards the PALS and a light zone harbouring centrocytes facing the red pulp are only seen in few cases (Fig. 6A). This is due to the fact that most of our specimens only contained decaying germinal centres with a rather symmetric structure. Similar to rats, the majority of the germinal centre B cells are surface $IgM^{+}IgD^{-}$ (Fig. 6C,D,F). The expression of IgM on germinal centre cells is, however, reduced in comparison to other B cells. In addition, IgM^{+} germinal centre B cells are difficult to distinguish immunohistologically from the FDCs carrying IgM immune complexes. At the outer circumference of the germinal centre towards the corona, and also inside some germinal centres, substantial numbers of $CD3^{+}CD4^{+}$ T lymphocytes occur (Fig. 7A). It is well-established that these cells play a decisive role for antigen-specific B lymphocyte differentiation to memory or plasma cells (see Sect. 5.1.3). Tingible body macrophages strongly positive for CD68 also form normal constituents of germinal centres (Fig. 6E).

Fig. 9. B cells in the human splenic white pulp and visualisation of the human perifollicular zone. **A** Expression of the CD20 B cell-surface antigen detected by MAb L26 reveals four distinct compartments in human secondary follicles. The germinal centre (*gc*) appears light due to the presence of unstained CD4-positive T cells and follicular dendritic cells. The corona (*co*) is populated by smaller cells and stains more darkly because of the nuclear counterstain. The marginal zone (*mz*) is densely populated by larger CD20-positive B cells with more cytoplasm and thus is perceived as paler. The perifollicular zone (*pfz*) in this particular specimen harbours many CD20-positive cells intermingled with monocytes and some neutrophils. The large unstained areas in this region represent capillary sheaths. Same specimen as in Fig. 6A. ABC technique on paraffin section using DAB chromogen. Bar = 80 μm

4.1.2.2
Corona (Mantle Zone)

The germinal centre is surrounded by a dark corona of variable thickness composed of small recirculating B lymphocytes. These cells strongly express IgM, IgD (Fig. 6C,D,F) and CD23, the low affinity receptor for IgE (Fc epsilon RII). The corona represents the equivalent of a primary follicle. Thus, a few single scattered T lymphocytes and CD68-positive macrophages are always present in this compartment (Fig. 6E).

4.1.2.3
Marginal Zone

The MZ predominantly harbours large pale IgM^+IgD^- B cells. In some specimens it is, however, apparent that the MZ B cells are not totally negative for IgD, but that they have a low level of expression (Fig. 6F). In addition, MZ B cells are negative for CD23. In contrast to rats, there is no distinct inner border of the MZ. In fact, the IgD^+ B cells of the corona and the IgD^- MZ B cells intermingle for some distance at the transition of both compartments. Staining with HE and detection of surface IgM and the CD20 and CD79a molecules in paraffin sections reveals that the MZ is subdivided into two parts in most (but not in all) follicles. There is a broader inner compartment formed by clear B cells, then following outside is a layer of unstained flat elongated cells representing the myofibroblasts described above, which is covered by a smaller outer layer of B cells. In most specimens, the B lymphocytes of this outer layer stain more intensely with HE and with antibodies against surface IgM, CD20 and CD79a than those of the inner layer (Figs. 6B, 9B). The outer layer of the MZ forms a small band of B cells which continues along the outer surface of the PALS forming the border to the red pulp (Fig. 9C).

The MZ represents a clear-cut B lymphocyte compartment in rats and humans. However, in humans variable numbers of $CD3^+CD4^+$ T lymphocytes may occur in the MZ. The number of T lymphocytes in the MZ is, however, not only variable among individuals but also among single follicles within the same specimen. Obviously, the numbers of $CD4^+$ T cells in the MZ change dynamically. This may depend on the kinetics of lymphocyte migration across the MZ. Interestingly, the T cells observed in the MZ tend to be associated with the asm-1-positive myofibroblasts (see Sect. 4.1.1), subdividing the MZ into an inner and an outer compartment. This phenomenon may

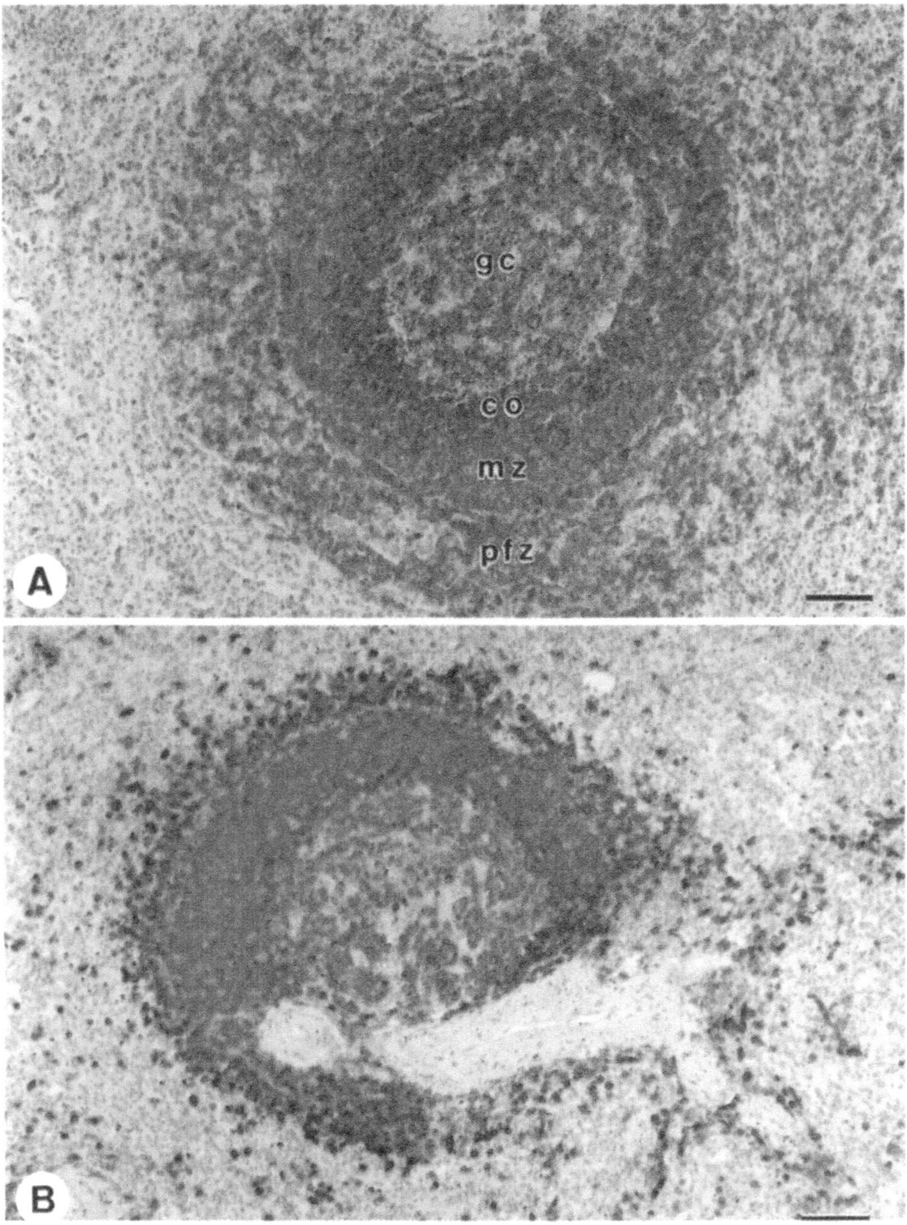

Fig. 9 *(continued)*. **B** Expression of the CD79a B cell-surface antigen detected by MAb JCB117 in a different specimen reveals another more frequent arrangement of B cells in secondary follicles. The germinal centre is rather light due to the presence of CD4-positive T cells and follicular dendritic cells. The corona is small and the marginal zone consists of two clearly visible compartments separated by a row of unstained myofibroblasts. The cells of the outer marginal zone are more intensely stained than those of the inner marginal zone. Same spleen as in Fig. 6F. ABC technique on autoclaved paraffin section using DAB chromogen. Bar = 80 μm

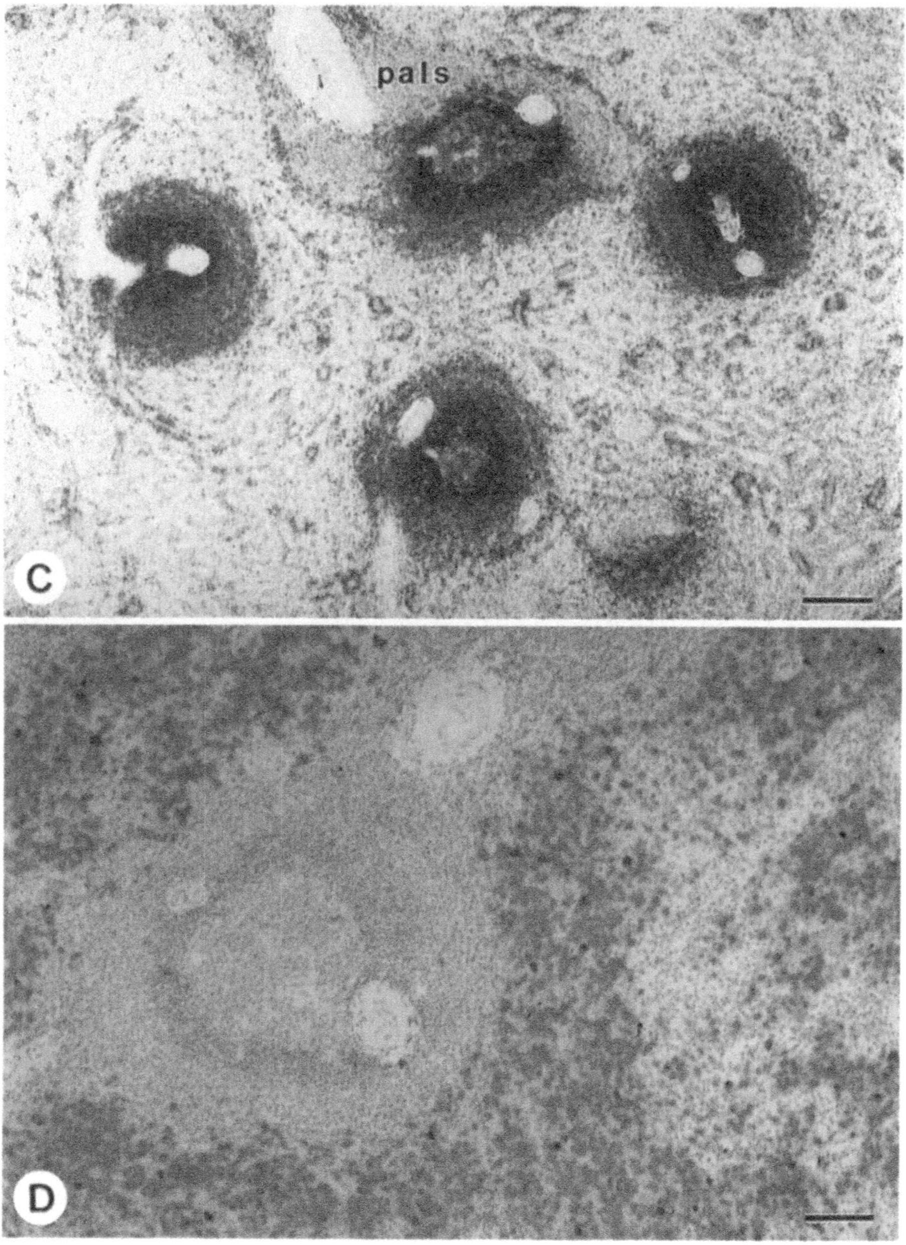

Fig. 9 *(continued)*. **C** Overview of the distribution of CD20-positive B cells stained by MAb L26. Follicles are the predominant component of the human white pulp. The B cells of the outer marginal zone continue at the periphery of the PALS. This phenomenon is, however, not encountered in all spleens. A large number of arterioles in the red pulp is surrounded by B cells. 35-year-old male patient with splenic trauma. ABC technique on paraffin section using DAB chromogen. Bar = 200 μm. **D** Expression of CD15 in the perifollicular zone demonstrated by MAb 28. Monocytes and granulocytes are positive. These cells accumulate outside the marginal zone. The staining in the red pulp (*right part* of figure) is less dense. ABC technique on cryostat section using DAB chromogen. Same organ as in Fig. 7F. Bar = 100 μm

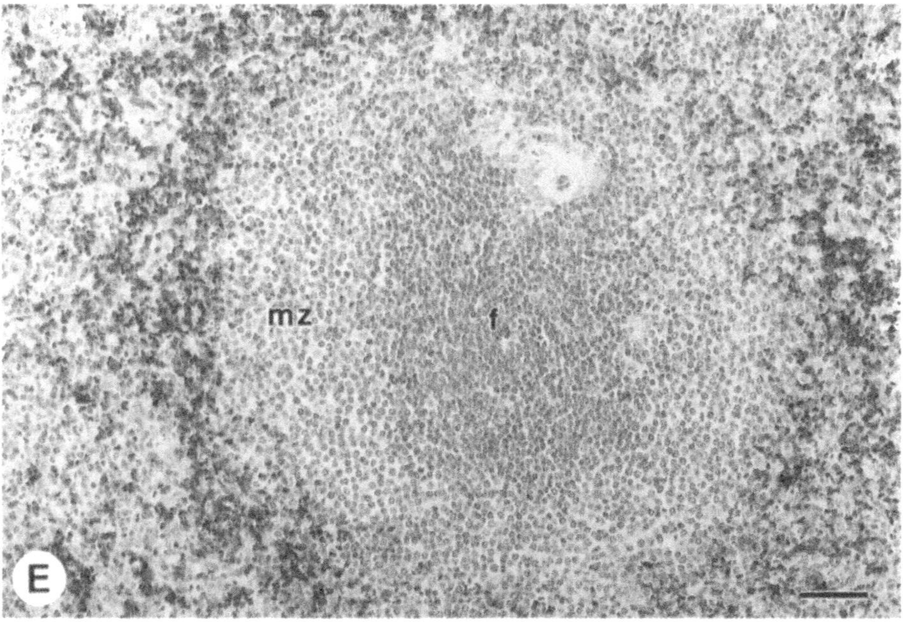

Fig. 9. E Accumulation of erythrocytes in the perifollicular zone revealed by peroxidase reaction with DAB. Erythrocytes are present in spaces outside the marginal zone (*mz*). *f* indicates primary follicle or tangentially sectioned corona. Same specimen as in Fig. 5B. Bar = 50 μm.

be due to the unique expression of adhesion molecules of the integrin family, extracellular matrix ligands or chemotactic factors by the myofibroblasts (see Sect. 4.1.1). T lymphocytes associated with the myofibroblast layer almost exclusively belong to the $CD4^+$ subset (Fig. 7A,B). Only a few single CD8 alpha chain-positive cells are found in the MZ and even less occur in the corona or germinal centre (Steiniger et al. unpublished). Instead, the highest density of CD8 alpha chain-positive cells is present in the red pulp and sometimes, but not always, also in the PALS (Fig. 7B).

HE-stained paraffin sections and antibodies against endothelial cell antigens, such as thrombomodulin and CD31, clearly show that neither a marginal sinus nor a perimarginal sinus occur around the human MZ. This is in contrast to the findings obtained with corrosion casts of human spleens (Schmidt et al. 1988, see below). Similar to all other follicular compartments, the MZ contains scattered CD68-positive macrophages. These MZMϕ express CD68 only weakly and they do not express sialoadhesin, as observed in rats. In addition, MMMϕ are absent in humans as demonstrated by lack of staining for acid phosphatase and sialoadhesin (Figs. 5B,D, Fig. 10A).

4.1.2.4
Perifollicular Zone

The surroundings of the follicle outside the MZ form a special region, which is not present in rats. This region was first described as the "perifollicular zone" by van Krieken and te Velde (1986, 1988). However, its presence was already noted by Weidenreich (1901) and by Herrlinger (1949) and called "Knötchenrandzone" by the former author. Other researchers have, however, neglected or incorrectly attributed this compartment, failing to distinguish it properly from the MZ (Saitoh et al. 1982; Brozman 1985a, 1985b).

The perifollicular zone is, indeed, somewhat difficult to assess, because it normally is a part of the red pulp adjacent to the follicle. It represents a kind of shell surrounding the MZ. With respect to lymphocytes, its composition resembles that of splenic red pulp cords. In certain conditions, however, the perifollicular zone can be occupied by B lymphocytes which are similar to the cells of the outer MZ. In this case, the perifollicular zone may as well be attributed to the white pulp and to the follicle, as it is difficult to distinguish from the outer MZ by B cell staining (Fig. 9A).

In most spleens of our study, the perifollicular zone is composed of small blood-filled spaces which are densely packed with erythrocytes and contain larger numbers of monocytes and granulocytes than other parts of the red pulp (Fig. 9D, E). The presence of monocytes and granulocytes is visualised by antibodies against CD14 and CD15 or by MAb KiM2. Staining for red cells by diaminobenzidine reaction demonstrates that the perifollicular region contains the most conspicuous accumulations of erythrocytes in the spleen (Fig. 9E). The erythrocytes are found in spaces without an endothelial lining, which lack thrombomodulin and CD31 (PECAM-1). Thus, the perifollicular zone is a part of the "open" circulation of the spleen, similar to the splenic cords. Staining for type IV collagen and for endothelial cells clearly reveals that the structure of the perifollicular blood-filled spaces differs from that of red pulp sinuses, although some sinuses begin or end in the perifollicular zone. The absence of endothelial cell antigens and collagen also excludes the existence of a special perimarginal sinus. There are, however, a few branched CD31-positive capillaries in the inner perifollicular zone which appear to continue towards the red pulp (Steiniger et al. unpublished).

The perifollicular zone contains very peculiar vessels, the sheathed capillaries. These vessels are absent in rats, but they are observed in humans when macrophages are visualised. Acid phosphatase activity or the CD68 antigen occur in typical macrophages in the red pulp and in the perifollicular region. In addition, both markers are weakly expressed in single cells within the marginal zone. Interestingly, in the perifollicular zone, acid phosphatase and CD68 are additionally present in cells which form the sheaths of capillaries. When the MAb HSN-1 or an antiserum against human sialoadhesin are applied to paraffin sections, the capillary sheaths are stained exclusively (Fig. 10A). With the exception of the immediate subcapsular region of the spleen, capillaries with sialoadhesin-positive sheaths are only present in the perifollicular zone (Fig. 10A). In isopropanol-fixed cryostat sections, MAb HSN-1 obviously detects more antigenic determinants than in paraffin sections and also weakly stains macrophages in the red pulp and macrophages in the perifollicular zone, in addition to the strongly reactive capillary sheaths. Sialoadhesin-positive cells are absent from the MZ also in cryostat sections. By double-staining for sialoadhesin and CD68 in paraffin sections all sialoadhesin-positive cells are positive for CD68. The sialoadhesin-positive capillary sheaths in the middle of the perifollicular zone appear to consist of two layers of macrophages, while the beginnings and the ends of the sheaths only comprise one layer. In paraffin sections the inner layer of macrophages close to the capillary endothelium seems to be sialoadhesin-negative (Fig. 10B). However, it cannot be excluded that this is due to low reactivity of the inner cells which is missed in paraffin sections. Class II MHC molecules (HLA-DP, -DQ and -DR) are absent from the sheath macrophages (Steiniger et al. unpublished). Interestingly, in some spleens strong sialoadhesin reactivity is only present in scattered branched macrophages in the perifollicular zone, but not in capillary sheaths. The most impressive example came from a patient with pancreatitis (Fig. 5B). We speculate that the sialoadhesin-positive macrophages can either leave their capillary sheaths under certain conditions to migrate into the perifollicular region or that they arrive via the perifollicular zone to reconstitute the sheaths. Interestingly, in two other spleens removed from patients with gastrointestinal malignancies, scattered strongly sialoadhesin-positive macrophages were also present in the perifollicular zone while the capillary sheaths were observed in a state of dissolution (Steiniger et al. unpublished, Fig. 10C).

The selective accumulation of sialoadhesin-positive cells, B lymphocytes and monocytes or granulocytes in the perifollicular zone indicates that this compartment may be a region where cells arrive via the blood to enter the splenic parenchyma (see Sect. 8.5). Thus, we reject the interpretation that the perifollicular erythrocytes in traumatised spleens simply represent "perifollicular haemorrhage" due to splenic injury (Farhi and Ashfaq 1996). In contrast to the results of Farhi and Ashfaq (1996) we also observed perifollicular accumulations of erythrocytes, monocytes and granulocytes in autopsy specimens, if autolysis permitted interpretation. Perifollicular erythrocyte accumulations have also been described by Millikin (1969) in autopsy cases. We cannot exclude that the conditions during removal of the spleen influence the composition of the perifollicular zone. Indeed, the accumulation of white cells is variable among different specimens. However, we assume that these variations reflect the dynamics of white cell adhesion and migration irrespective of organ lesions (see Sect. 8.5).

According to our opinion, the spaces of the perifollicular zone have led to an erroneous interpretation of corrosion casts by Schmidt et al. (1988, 1993) and Yamamoto et al. (1979). These authors suggest that a marginal sinus and a perimarginal sinus are present around human splenic follicles. It is, however, more likely, that the accumulations of casting material represent filling of the perifollicular zone, which went undetected because the follicle and its lymphocytes were digested away afterwards. These methodological limitations obviously also led Fujita (1982) and Kashimura and Fujita (1987) to assume filling of the "marginal zone" with casting material. With the exception of investigations using corrosion casts, most evidence is against the existence of a marginal sinus in humans, including several scanning electron microscopic studies (Saitoh et al. 1982; Sasou et al. 1986).

Our results must be discussed within the context of the few immunohistological studies on normal human spleens in order to clarify some apparently controversial findings. A special activated phenotype of MZ B cells and the occurrence of T lymphocytes, granulocytes and monocytes were described by Hsu et al. (1983, 1985) in the MZ. These findings have led to confusion in several histology textbooks. In our opinion, an

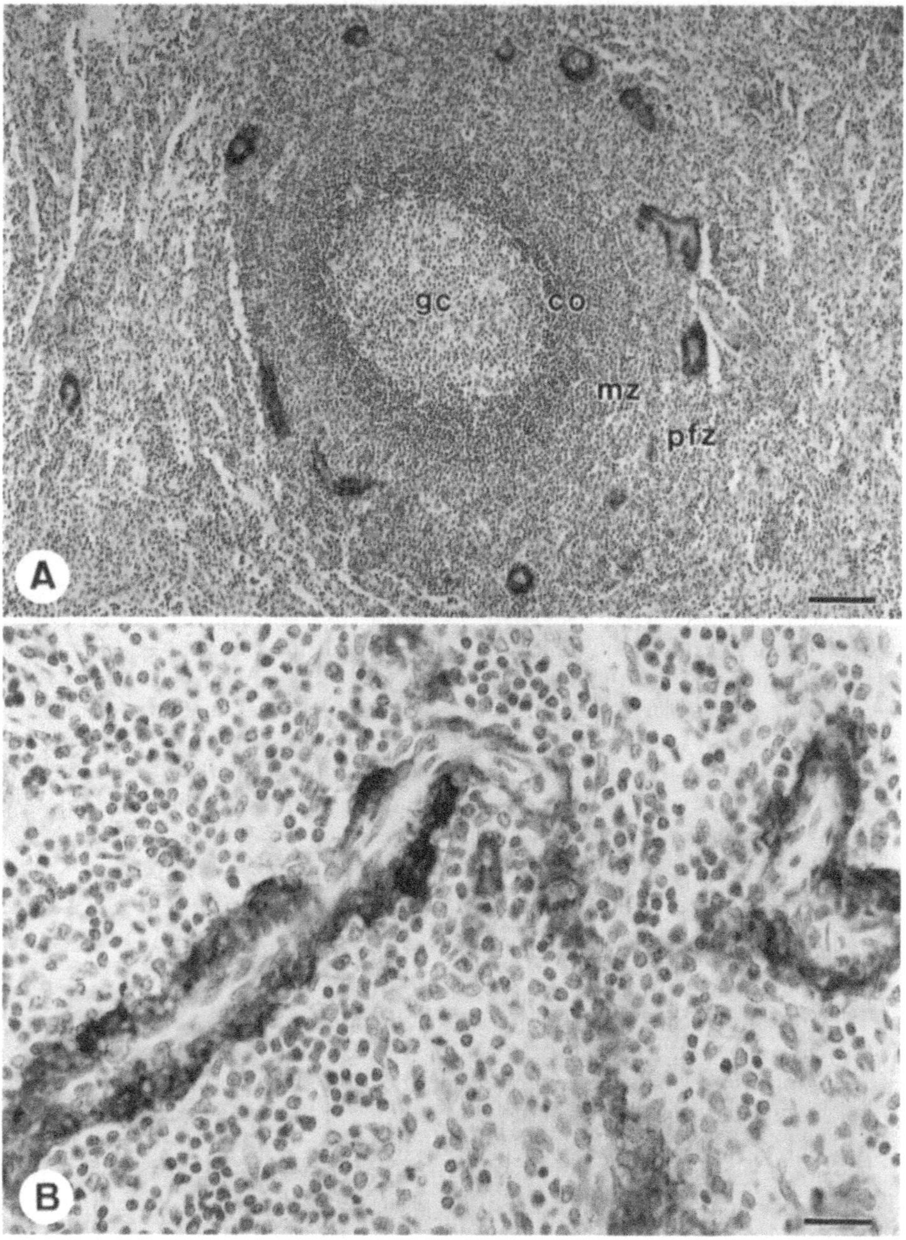

Fig. 10. Sheathed capillaries in the human perifollicular zone and B cells in the human red pulp. **A** Sheathed capillaries demonstrated in the perifollicular zone (*pfz*) by strong staining with MAb HSN1 against human sialoadhesin. Sialoadhesin-positive capillary sheaths are closely associated with primary and seconday follicles and occur outside the marginal zone (*mz*). *gc*, germinal centre; *co*, corona. Same spleen as in Fig. 6A. Bar = 100 μm. **B** Branching capillary sheaths stained for sialoadhesin (MAb HSN1) at higher magnification. The endothelium and the innermost macrophage layer appear sialoadhesin-negative. Same spleen as in Fig. 6A. Bar = 25 μm. ABC technique on paraffin sections using DAB chromogen (A,B).

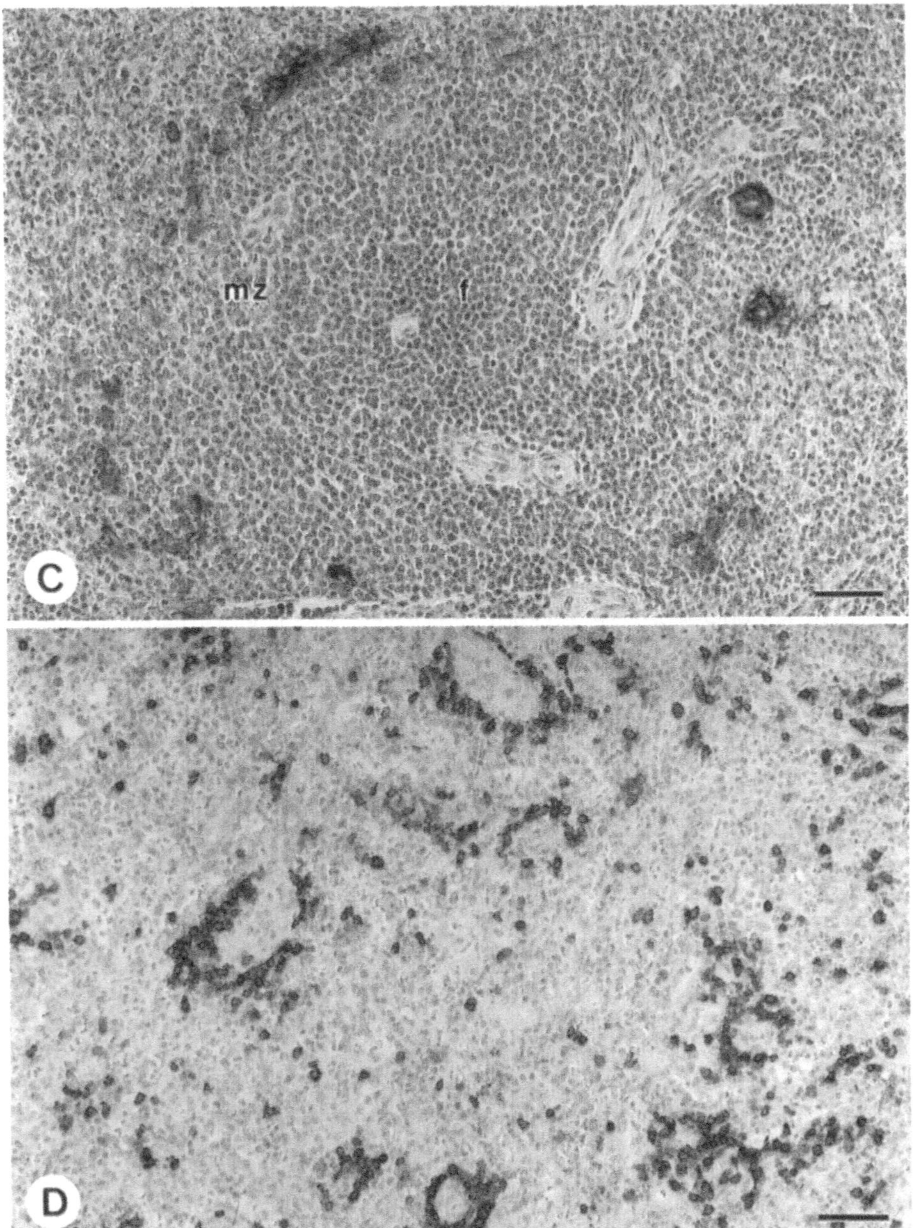

Fig. 10 *(continued)*. **C** Dissolving capillary sheaths in the spleen of a 55-year-old male with splenic trauma. The pathologic diagnosis was splenitis. In this specimen two intact capillary sheaths are present to the right of the follicle (*f*), while on the left side single stellate sialoadhesin-positive cells are distributed in the perifollicular zone outside the marginal zone (*mz*). MAb HSN1 demonstrated by ABC technique on paraffin section using DAB chromogen. **D** CD79a-positive B cells in the human splenic red pulp visualised by MAb JCB117. The red pulp contains a large number of arterioles, which are surrounded by thin accumulations of B lymphocytes. 65-year-old female with colon carcinoma. ABC technique on autoclaved paraffin section using DAB chromogen. Bars = 50 μm (C,D).

incorrect attribution of the single zones of secondary follicles is responsible for the problem. The apparent controversies may be solved by considering three facts:

1. The MZ is a memory B cell compartment. It is also traversed by recirculating IgD^+ B cells and certain T cells.
2. $CD4^+$ T cells are not always prominent in the MZ, but accumulations of migrating $CD4^+$ T cells may be visible in this compartment under certain conditions.
3. Monocytes and granulocytes are definitively located outside the MZ in the perifollicular zone.

Thus, it is very likely, that the IL-2 receptor alpha chain is not expressed in MZ B cells as reported by Hsu et al. (1983, 1985), but in MZ T cells. The conclusion that 40%–50% of the MZ lymphocytes are T cells (Hsu et al. 1983) was obviously provoked by the small number of organs investigated. It does not reflect the predominant situation. We agree with Timens et al. (1989) that the B cells of the MZ do not represent an activated population. Our findings are in accordance with those of Timens and Poppema (1985), Grogan et al. (1983, 1984) and van Krieken and te Velde (1986). In fact, Timens and Poppema (1985) were the first to describe $CD4^+$ T cells located in a ring-like fashion within the MZ. Timens and Poppema (1985) and Grogan et al. (1983) found a difference in the distribution of $CD4^+$ and $CD8^+$ T cells in the spleen. Van Krieken and te Velde (1986) are the only authors who give an immunohistological definition of the perifollicular zone. Buckley et al. (1987, 1991) have assessed the distribution of subsets of macrophages and dendritic cells in the human spleen. In addition, the authors show that cells with a monocyte or granulocyte phenotype accumulate around follicles. The findings of Tanaka et al. (1984) also correspond to our results. Interestingly, these authors published a schematic drawing of the follicular structure in humans which is very similar to Fig. 8, showing the correct number of zones. However, Tanaka et al. (1984) did not realise that the outermost surroundings of the follicle represent the perifollicular zone and not the MZ.

4.2 Red Pulp and Splenic Vessels

In humans, the transitions from the MZ to the perifollicular zone and from the latter to the splenic red pulp proper are gradual without precise borders. Obviously, all white pulp compartments are dynamic and can be either expanded or contracted due to the velocity, quantity and preferential direction of migrating lymphocyte subpopulations.

Similar to rats, the red pulp of the human spleen consists of sinuses and cords. The sinuses originate in the perifollicular zone and finally open into trabecular veins draining the organ. There may be venules forming a connection between sinuses and trabecular veins, but these vessels are extremely short. Human splenic sinuses have a specialised endothelium, which differs from all other endothelia in the spleen. Our own results show that sinus endothelia express the alpha chain of CD8 and thrombomodulin while being only weakly positive for CD31. Sinusoidal CD8 expression has been described by all authors investigating this molecule in the human spleen (Stuart and Warford 1983; Giorno 1984; van der Valk et al. 1984; Timens and Poppema 1985; Van Krieken and te Velde 1986; Buckley 1991). The function of CD8 in sinus endothelia is enigmatic. Timens and Poppema (1985) indicate that CD8 positivity of sinuses is acquired during ontogeny and is absent in neonatal spleens. Buckley et al. (1985, 1991) and Giorno (1984) have also demonstrated, among other antigens, CD36 (OKM5), lysozyme, factor VIII, MHC class II molecules, non-specific esterases and the transferrin receptor (CD71) in sinus endothelia. Thus, according to Buckley (1991), the cells exhibit a peculiar mixture of endothelial and phagocytic characteristics.

It is generally accepted that sinusoidal endothelial cells do not produce a continuous basal membrane, but are attached to ring fibres of basement membrane-like material on their outside. This necessitates a peculiar arrangement of microfilaments beneath their cell membrane (Drenckhahn and Wagner 1986). Whether sinus endothelia are contractile in humans and whether the inter-endothelial slits observed in scanning electron micrographs represent permanent openings, a manifestation of contractility or a pure artefact, has not been clarified (Chen and Weiss 1972; Fujita 1974). Intravital video microscopy, transmission and scanning electron microscopy, in humans and rodents, unequivocally indicate that erythrocytes enter splenic sinuses from the outside, i.e. from the splenic cords, via these slits (see Sect. 3.2). Whether this is also true for leukocytes, has not been investigated in detail up to now. Thus, a minor portion of splenic perfusion is obviously directed from the "open" cord spaces to the "closed" part of the vasculature, namely the sinuses.

Apart from splenic cords and sinuses, there is a third rather prominent structure in the human splenic red pulp, the arterioles. Similar to rats, these vessels are covered by a thin sheath of B lymphocytes. Thus, B cells are clustered in the red pulp, while T cells are evenly distributed (Fig. 10D). These rather large arteriolar B cell sheaths may correspond to the surroundings of "terminal arterioles" mentioned by some authors as locations for antibody-forming B cells in rodent spleens during primary immune reactions (see Sect. 5.1.2). It may well be that in humans these structures have in the past been mistaken for "ellipsoids", i.e. large capillary sheaths. In birds, cats, dogs or swine, large elliptic-shaped sheaths consisting of macrophages and fibroblasts normally surround vessels with rather leaky walls in the red pulp (Snook 1950). In humans, sialoadhesin-positive sheathed capillaries are only found in the perifollicular zone and true "ellipsoids" do not occur. Whether sialoadhesin-negative macrophage sheaths are additionally present in the red pulp remains to be investigated.

In some human spleens penicillar arterioles are observed, i.e. a bunch of arterioles arises from a common larger vessel in the red pulp. However, this is the case in only a minority of specimens investigated. The absence of penicillar arterioles in most of the spleens in our study is enigmatic and might be due to the location of the individual specimens within the organ. However, the distance of the specimen from the organ surface does not seem to play a role, because penicillar arterioles were either present or absent in material containing the splenic capsule. Van Krieken et al. (1988) also note variations in the occurrence of penicillar arterioles. These authors speculate that an overall reduction of splenic lymphocytes with increasing age causes the penicillar vessel branches to approach one another and to become visible. This explanation is in accordance with our observation that penicillar vessels are not always associated with a prominent PALS. Thus, the conclusion is that only a minority of human splenic vessels form penicillar arterioles and sheathed capillaries.

4.3
Summary

We demonstrate by published (Steiniger et al. 1997) and new unpublished evidence that the human splenic white pulp differs from that of rats in the following aspects (Figs. 1,8):

1. A secondary follicle comprises up to four zones, the germinal centre, the corona, the MZ and the perifollicular zone.
2. The MZ is primarily visible around follicles. It forms a regular part of these structures while it is almost lacking around the T cell region.
3. A marginal sinus and MMMφ do not occur. Thus, the PALS or corona are not clearly separated from the MZ.
4. An additional region is found around the follicles, the "perifollicular zone". This zone may be regarded either as a part of the red pulp or of the follicles.
5. MZMφ do not express sialoadhesin.
6. Capillary macrophage sheaths and, occasionally, macrophages distributed in the perifollicular zone outside the MZ strongly stain for sialoadhesin.

5 Function of Splenic Compartments

Our knowledge about the immunological functions of different splenic compartments is almost entirely based on observations in mice and rats. As the basic principles of cooperation among immunocompetent cells appear to be relatively similar in different mammalian species, extrapolations to humans are justified. However, it has to be expected that human spleens will also exhibit species-specific and thus far unknown peculiarities with respect to the spatial distribution of certain functions.

Antibody production has been the splenic function most thoroughly investigated up to now. The spleen produces antibodies against so-called "T cell-dependent" as well as against "T cell-independent" antigens. The spleen does, however, differ from other secondary lymphoid organs[1], because there is a special compartment, the MZ, where IgM-positive B memory cells[2] for a certain type of T cell-independent antigens, the "T independent type 2 (TI-2) antigens", localise selectively. TI-2 antigens comprise polymeric polysaccharides, glycoproteins or glycolipids with a large number of repetitive B cell epitopes and a long half-life in vivo, such as occur in bacterial capsules or in the envelopes of certain viruses (Mond et al. 1995; Bachmann et al. 1993). The MZ plays an essential role during the initial rapid IgM antibody response especially to the TI-2 neutral polysaccharides of bacterial capsules (Amlot and Hayes 1985; Hosea et al. 1981). Thus, in humans and rodents, the MZ is associated with a life-sustaining function of the spleen during the initial phase of systemic infection by encapsulated bacteria or certain viruses.

In addition to this special function (see previous footnotes), the spleen, like any other lymphoid organ, also contributes to the production of antibodies against protein antigens. Memory cells against protein antigens also localise to the MZ. B cell reactions against proteins are, however, totally dependent on initial "help" provided by T lymphocytes. With respect to antibody production, immunologists use the term "help" operationally to summarise several types of interaction among T cells and B cells, which support the activation of B "effector" lymphocytes when specific antigen is recognised by their cell surface Ig. In general, "help" circumscribes the fact that, in addition to the ligation of the antigen-specific surface receptors, a so-called "second

[1] The difference from other secondary lymphoid organs may be more quantitative than qualitative, because in humans, B cells with an MZ phenotype may also occur around follicles in lymph nodes and other secondary lymphatic organs (own unpublished observations; van Krieken et al. 1989; Spencer et al. 1998).

[2] It has, however, not been shown unequivocally that all MZ B cells specific for TI-2 antigens are memory cells.

signal" is necessary to activate the effector function of a B or T lymphocyte. This signal may comprise the interaction of "co-stimulatory" surface molecules (see below) occurring on the helper cell (or on an antigen-presenting cell) and on the effector cell and/or the secretion and binding of soluble mediators, such as certain interleukins. Binding of antigen to its specific cell surface receptor (Ig in B cells and the TCR in T cells) in the absence of the second signal is supposed to lead to apoptotic death of the effector lymphocyte and may thus provide a safeguard against autoimmunity (Matzinger 1994).

In rodents, the cellular interactions during antibody responses in the spleen have been primarily investigated using modified protein antigens, so-called hapten-carrier systems. We shall thus first review the function of different splenic compartments during helper T cell-dependent B cell reactions to protein antigens before briefly discussing T cell-independent antibody production. We are aware of the fact that our presentation is somewhat simplistic with respect to the role of myeloid dendritic cells in T cell-dependent antibody reactions, implying that these dendritic cells have a constitutive capacity for antigen presentation in the mature state (Lanzavecchia 1998). In addition, the role of MHC class II molecules in antigen presentation is more sophisticated than can be outlined in this review (Lechler et al. 1996; Germain et al. 1996).

In the murine spleen, primary B cell reactions against protein antigens can be split into a minimum of four consecutive phases occurring in different compartments, which are summarised in Sect. 5.1.1–5.1.4 below.

5.1 Splenic White Pulp Compartments during Primary T Cell Dependent Antibody Responses against Protein Antigens

5.1.1 Priming of $CD4^+$ Helper T Cells by Dendritic Cells in the PALS

T cell-dependent primary antibody responses necessitate the activation of naive $CD4^+$ helper T cells by dendritic cells, which are located in the PALS or which migrate to the PALS (Steinman et al. 1997). Up to now, class II MHC-positive IDCs have been thought to be responsible for T cell activation in the PALS, but this is still controversial (see below). Dendritic cells are the only antigen-presenting cells capable of stimulating naive T cells (Matzinger 1994). Their function is to endocytose antigenic proteins, degrade (or process) them proteolytically and to finally „present" protein fragments at their cell surface.

The decisive final step of "antigen presentation" implies the display of each antigenic peptide associated with a class II MHC molecule on the surface of the dendritic cell. Class II MHC molecules are capable of binding antigenic peptides and the complex of both, class II MHC molecule and peptide, forms the ligand for the antigen-specific T cell surface receptor (TCR) of $CD4^+$ helper T cells. This phenomenon is called the "MHC class II restriction" of the helper T cells. The MHC-bound peptide or peptide surrogate is not only essential for antigen recognition by T cells, but it also serves to stabilise the conformation of the MHC molecule after its synthesis in the endoplasmic reticulum and also after its insertion into the cell membrane (Germain et al. 1996).

After binding of the TCRs to MHC molecules plus peptides on an antigen-presenting cell, two types of cell interaction molecules must be additionally engaged for proper activation of naive helper T cells: co-receptors and co-stimulatory molecules. The decisive co-receptors on MHC class II-restricted T cells are the CD4 molecules which bind to a nonpolymorphic part of the alpha and beta chains (most likely alpha 2 and beta 2 domains) of class II MHC molecules on the antigen-presenting cell (König et al. 1996; Li et al. 1998; for review of co-receptors see Zamoyska 1998)[1]. Proper activation of a $CD4^+$ T lymphocyte necessitates a second signal, called co-stimulation, in addition to engagement of the CD4 co-receptors. The most important molecules for co-stimulation of naive T cells by antigen-presenting cells are two membrane receptor-ligand pairs: the CD28 molecule on T cells binding the CD80 or CD86 molecules on the presenting cell and the CD40 ligand on T cells binding the CD40 molecule on the presenting cell. CD28, CD80 and CD86 belong to the immunoglobulin protein superfamily, while CD40 ligand (CD154) and CD40 are members of the tumour necrosis factor (TNF) and the TNF-receptor family, respectively. CD28 is constitutively expressed on T cells and recognises CD80 (also called B7-1) or CD86 (also called B7-2) on the surface of antigen-presenting cells (Lanier et al. 1995), while CD40 ligand is induced on T cells after TCR and CD4 occupancy and engages CD40 on the presenting cell. There is a mutual activation of T cells and antigen-presenting cells with the help of these co-stimulatory molecules. Thus, the interaction of one receptor/ligand pair may induce expression of the other, i.e. CD40 ligand binding to CD40 on an antigen-presenting cell upregulates CD80 and CD86 on the latter (Lenschow et al. 1996; Grewal and Flavell 1996; van Gool et al. 1996; van Kooten and Banchereau 1997; Ridge et al. 1998). Finally, signals for cytokine secretion and proliferation are transmitted to the T cell interior and provide the precondition for T cell help. The co-stimulatory molecules mentioned are of decisive importance for activation of T cells during primary T cell-dependent B cell reactions in the spleen. However, they are not exclusive and alternative interaction molecules can provide co-stimulation when the T cells or the antigen-presenting cells are not naive, but have been stimulated before (Wu et al. 1998).

Dendritic cells form a bone marrow-derived population of heterogeneous lineage, which is characterised by branched shape, very high expression of class I and class II MHC molecules in the mature state and almost ubiquitous distribution in the body (Vremec et al. 1992; Steinman et al. 1997; Banchereau and Steinman 1998). For effective antigen presentation dendritic cells have to complete several maturation steps. Immature dendritic cells are actively taking up protein antigens by receptor-mediated endocytosis, macropinocytosis and even phagocytosis, while their T cell activating potential and cell surface MHC class II molecule expression levels are rather low (Sallusto et al. 1995). More mature cells of this lineage have a reduced capacity for antigen uptake and proteolytic processing but they are very efficient at antigen presentation to specific T cells (Steinman and Swanson 1995; Pierre et al. 1997). This is in part due to increased expression and half-life of class II MHC molecules on their

[1] The second T lymphocyte subset, the cytotoxic T cells, use CD8 molecules as co-receptors for binding to class I MHC molecules on antigen-presenting cells.

surfaces (Cella et al. 1997). B lymphocytes, which constitutively express class II MHC molecules, are also capable of antigen presentation. Both B cells and dendritic cells load the fragments of exogenous or endogenous proteins, which they have processed proteolytically, onto newly formed MHC class II molecules. The loading occurs inside a special intracellular compartment corresponding to late endosomes, which harbours the newly synthesised MHC class II molecules in its membrane. This compartment has been called the „MHC class II compartment" (MIIC). In the MIIC, an appropriate peptide is bound to a special peptide-binding groove at the surface of the MHC class II molecule by replacing the peptide remnants of the so-called „MHC class II invariant chain" (Pieters 1997; Geuze 1998). The MIIC/endosomes are then transported to the cell surface, their membranes are fused with the cell membrane and the peptide-carrying MHC class II molecules are exposed to the exterior for antigen presentation to $CD4^+$ T cells (Geuze 1998).

Depending on their lineage, dendritic cells perform different functions. In mice, current classification distinguishes two to three lineages of dendritic cells, myeloid dendritic cells, lymphoid dendritic cells and Langerhans cells (the latter are, however, not always regarded as a separate dendritic cell population). Myeloid dendritic cells are distributed from the bone marrow to the organs by the blood involving a monocyte-like intermediate stage. The confrontation with different cytokine milieus in the respective organ decides whether the cell turns into a typical dendritic cell or into a macrophage (Peters et al. 1996). The lymphoid dendritic cell is supposed to share a lineage with T lymphocytes, B cells and natural killer (NK) cells in its most immature state (Saunders et al. 1996; Ardavin 1997). In the mouse spleen this cell type is characterised by expression of MHC class II and co-stimulatory molecules in combination with the alpha chain of CD8 surface molecules. This is astonishing because CD8 molecules in their alpha/beta chain configuration have so far been primarily regarded as co-receptors of thymus-derived MHC class I restricted cytotoxic T lymphocytes and, in the alpha/alpha configuration, of NK cells (Torres-Nagel et al. 1992). In contrast to myeloid dendritic cells, the $CD8^+$ lymphoid dendritic cells do not stimulate but delete antigen-specific $CD4^+$ T lymphocytes by inducing apoptosis (Süss and Shortman 1996; Kronin et al. 1997). This process is especially evident in the thymic medulla and might contribute to preventing T lymphocyte autoreactivity (Vremec et al. 1992). A negative immunoregulatory effect of dendritic cells might also be necessary at other sites, because it has been shown that dendritic cells in spleen and lymph nodes carry high amounts of self-peptides bound to their MHC class II surface molecules (Inaba et al. 1997). Thus, in mice MHC class II-positive dendritic cells belong to different populations with contrasting functions. It is not clear up to now which lineage of dendritic cells the so-called IDCs of the splenic PALS belong to. IDCs may represent lymphoid dendritic cells, which are derived from immature class II MHC^- $CD11c^+$ precursors located in the outer PALS or the MZ. These precursors may have arrived from the bone marrow or from other sites via the blood. They are supposed to migrate to the inner parts of the PALS on maturation and to acquire MHC class II molecules (Reis e Sousa et al. 1997). Steinman et al. (1997) suppose that the majority of dendritic cells, which have been recovered by digestion of splenic tissue for investigation of dendritic cell function in their classical in vitro experiments, may correspond to the immature dendritic cells of the outer PALS. These cells need maturation in vitro to become effective antigen-presenting cells for T lymphocytes.

5.1.1.1
Summary

The prerequisite of a B cell immune reaction against protein antigens in the spleen is the antigen-specific activation of $CD4^+$ T helper cells in the PALS. This is accomplished by dendritic cells, which exhibit MHC class II molecules at their cell surface. An MHC class II molecule serves to bind an antigenic peptide in a peptide-binding groove. The walls of the groove and the bound peptide are recognised by the antigen-specific receptor on the surface of a T helper cell. Another part of the MHC class II molecule binds to the CD4 co-receptor on the T cell. Additional ligation of co-stimulatory molecules on the antigen-specific T cell finally leads to activation and proliferation signals being transmitted into the T cell interior. There are several types of dendritic cells in the mouse spleen, supporting or inhibiting T cell activation. It is not clear, which dendritic cell type the IDCs of the PALS belong to.

5.1.2
Interaction of Primed $CD4^+$ T Cells with Antigen-Specific B Cells in the PALS and Formation of Extrafollicular Foci

$CD4^+$ T lymphocytes which have been specifically activated by dendritic cells subsequently move to the oPALS. At this site they meet activated antigen-specific B lymphocytes which arrive via the blood and the MZ. These B cells have bound and internalised their specific antigen through their Ig surface receptors. After processing of the antigen by B lymphocytes, antigenic peptides are presented at the B lymphocyte surface in the context of MHC class II molecules. Thus, B lymphocytes are capable of antigen presentation due to their constitutive expression of class II MHC molecules. However, in contrast to dendritic cells, B lymphocytes only stimulate primed, but not virgin, T cells. This is at least in part due to the pattern of co-stimulatory molecules at the B cell surface, which differs from that of dendritic cells. The antigen-specific interaction of $CD4^+$ T helper cells primed by dendritic cells and antigen-presenting blood-borne B lymphocytes leads to arrest of the B cells in the oPALS. The B cells receive further stimulation from the $CD4^+$ T cells and differentiate to plasmablasts and "plasma cells". The interaction between T cells and B cells leads to a mutual activation, demonstrated also by the fact that in mice T cells in the oPALS have a more activated phenotype than in the inner PALS (Gulbranson-Judge and MacLennan 1996). The CD40 molecule expressed on B cells and its ligand, which is induced on activated T cells, appear to exert a crucial role during stimulation and differentiation of B cells in the PALS (Gray et al. 1994).

Similar to the interaction of T cells and dendritic cells, which may be controlled for autoreactivity by lymphoid dendritic cells, the interaction of $CD4^+$ T cells and B lymphocytes in the oPALS is a complex event including controls. It has been shown, that B cells with high receptor occupancy due to autoantigen binding may be anergized or deleted by interaction with T cells, e.g. via the apoptosis-inducing Fas-Fas ligand system. Thus, in an autoantigen model in transgenic mice, the amount of surface immunoglobulin receptor ligation by autoantigen determines whether B cells are arrested in the oPALS and die, or whether they pass on into the follicles (Cook et al.

1997). Another explanation for the arrest of autoantigen-binding B cells in the oPALS is the exclusion of these cells from follicles due to competition with recirculating non-activated B cells (Cyster et al. 1994).

In case of optimal T cell help, the stimulated B cells turn into plasmablasts. In mice, the proliferating antigen-specific B cells form so-called "extrafollicular foci" in the oPALS at the border to the red pulp and around the so-called terminal arterioles. These vessels might correspond to MZ-bridging channels in rats (Jacob et al. 1991a; Jacob and Kelsoe 1992; Gulbranson-Judge and MacLennan 1996). The extrafollicular foci give rise to short-lived plasma cells with a half-life of about 3 days, which subsequently migrate to the cords of the splenic red pulp (Smith et al. 1996; MacLennan et al. 1997). The Ig secreted by these plasma cells is of low affinity for the antigen, because the cells have not passed germinal centres. Thus, affinity maturation due to hypermutation of the variable region genes of immunoglobulins (see Sect. 5.1.2) and subsequent selection of B lymphocytes has not yet occurred. In mice, the initial antibody response is of the IgM type and it is subsequently switched to IgG after about a week post-immunisation (Jacob et al. 1991a). Interestingly, a substantial number of B lymphocytes already switch to other Ig classes by somatic recombination at this stage. (Toellner et al. 1996). After a single application of protein antigen, the antibody-secreting cells of the extrafollicular foci start to vanish after day 10 due to apoptosis. This process is interpreted as a control for the propagation of autoreactive B lymphocytes, which have not been selected in a germinal centre (Jacob et al. 1992; Smith et al. 1996). The presence of Ig produced by the B lymphocytes, which have differentiated extrafollicularly, forms an essential prerequisite for the subsequent development of germinal centres.

5.1.2.1
Summary

The second step during a splenic B cell reaction against a protein antigen in mice involves the encounter in the oPALS of activated $CD4^+$ T cells with B cells which have internalised the specific protein antigen with the help of their surface Ig. These B cells present antigenic peptides to the TCR of $CD4^+$ T cells. The ligation of CD4 and of co-stimulatory molecules induces further mutual activation of both cell types which finally leads to differentiation of the B cells to plasmablasts and so-called plasma cells. The differentiated B cells accumulate in extrafollicular foci of Ig-producing cells in the oPALS. The Ig production by these B cells is of short duration and the Ig has only low affinity for the antigen. The formation of germinal centres depends on the presence of circulating Ig produced by the extrafollicular B cells.

5.1.3
Formation of Germinal Centres

There is a general consensus that prolonged as opposed to short IgG antibody responses to protein antigens necessitate passage of antigen-specific B cells through germinal centres at some stage of their differentiation (Fang et al. 1998). Germinal centres only arise when the low-affinity antibodies secreted by the antibody-producing cells of the extrafollicular foci and by plasma cells at other locations have bound their respective antigen and have formed immune complexes. These complexes finally localise on the surface of FDCs in primary follicles (Tew et al. 1997). It is not quite clear how the immune complexes reach the FDCs in the spleen, i.e. whether they arrive by the blood or by special antigen-transporting cells immigrating into the follicles, similar to those shown in lymph nodes (Nossal et al. 1966; Tew et al. 1989; Szakal et al. 1992). FDCs capture immune complexes with the help of Fc and complement receptors and exhibit the complexed antigen at their surface for very long periods (Yoshida et al. 1994; Tew et al. 1997; Fang et al. 1998). The antigen-antibody complexes are retained for months and even for years (Mandel et al. 1980). How this long-term preservation is achieved remains an enigma. There is also no information as to whether the Ig involved in the complexes is totally shielded from the surroundings or whether it is slowly replaced by antibodies of higher affinity for the antigen. FDCs are supposed to be resident cells differentiated from the local connective tissue. Some controversy still prevails about this fact, because it has not been totally excluded that FDCs only immigrate from the bone marrow in the very early pre- or post-natal period and remain stationary afterwards (Yoshida et al. 1995; Tew et al. 1997; Matsumoto et al. 1997a; Tkachuk et al. 1998).

Antigen-antibody complexes localised on FDCs and the immigration of antigen-specific $CD4^+$ T lymphocytes into the follicles are both essential for the formation of germinal centres. The $CD4^+$ T cells in germinal centres are derived from those cells which have been activated first by IDCs and then by antigen-presenting B lymphocytes in the oPALS, as assessed by MAbs against certain families of the variable regions of antigen-specific TCRs (Gulbranson-Judge and MacLennan 1996). In general, primary follicles in lymphatic organs serve as recirculation compartments for mature B lymphocytes. They represent locations where primed B cells survey FDCs for cognate antigen and where they receive stimuli to continue recirculation between the blood and the secondary lymphatic organs if the relevant antigen is absent. If, however, the cognate antigen is encountered in the immune complexes residing on FDCs, the specific B cells stay in the follicle, proliferate and hypermutate. Subsequently, B cells with improved binding affinity of their surface Ig are selected (see below). Thus, specific B cells form germinal centres, which arise on days 3 to 4 (Jacob and Kelsoe 1992; Gulbranson-Judge and MacLennan 1996) of a primary B cell reaction when sufficient T cell help is present. During a secondary reaction, the speed of germinal centre development is increased. The life-span of a germinal centre depends on the amount of antigen available and can amount to weeks or months in systems where antigen persists in the body (Bachmann et al. 1996). However, after a single i.p. injection of hapten-conjugated carrier protein in alum, antigen soon becomes limiting and the germinal centre response does not last longer than 20–30 days (Kosco-Vilbois et al. 1997; Takahashi et al. 1998).

Liu et al. (1991) have divided the germinal centre reaction into three successive stages. First, B cell blasts accumulate, then a germinal centre with a dark and a light zone of differentiating B cells is established and, finally, the germinal centre is again transformed into a non-structured accumulation of B cell blasts persisting for a prolonged period. The B cells "founding" a germinal centre belong to a small number of about three to six different clones, i.e. germinal centres are derived oligoclonally from precursor B cells expressing IgD and IgM at their surface (Kroese et al. 1987; Jacob et al. 1991a; Liu et al. 1991; MacLennan et al. 1997). The first phase of the germinal centre reaction is characterised by antigen-specific B cell blasts accumulating among the FDC network. Then a full-blown germinal centre develops, consisting of a dark zone and a light zone. Both zones are often arranged asymmetrically and the dark zone is always oriented towards the T cell compartment. The germinal centre is surrounded by the corona or mantle zone of small recirculating B lymphocytes, which are not attracted into the germinal centre for lack of relevant antigens on FDCs.

B cells that encounter cognate antigen first enter the dark zone of the follicle, where they proliferate and turn into centroblasts (MacLennan et al. 1997). During this proliferation the genes encoding the variable antigen-binding portions of the cell surface Ig of centroblasts undergo a process of somatic hypermutation (Jacob et al. 1991b, Küppers et al. 1993). A portion of these Ig-hypermutated cells is obviously immediately abolished by apoptosis. However, centroblasts are generally prone to apoptosis, even before the hypermutation process starts (Lebecque et al. 1997). The surviving centroblasts continue to differentiate into centrocytes which do not divide (Liu et al. 1991) and which form the predominant cell type in the light zone of the germinal centre. The light zone is supported by a crescent-shaped network of FDCs bearing high amounts of immune complexes. During their movement through this compartment the centrocytes are selected according to the affinity of their mutated Ig for the antigen retained by FDCs. This selection mechanism operates in a rather intricate way: it is supposed that FDCs shed parts of their membranes containing bound immune complexes (so-called "iccosomes" or immune-complex-coated bodies) and that these membrane particles are taken up by centrocytes via their surface Ig (Szakal et al.1988). Centrocytes then present antigenic peptides derived from the internalised immune complexes in association with their MHC class II molecules to the primed antigen-specific $CD4^+$ T cells, which have immigrated into the germinal centre from the oPALS (Gulbranson-Judge and MacLennan 1996). This process leads to stimulation and proliferation of the antigen-specific germinal centre T cells. It is not quite clear whether a special population of germinal centre dendritic cells, which are similar to IDCs, may also be necessary for this process (Grouard et al. 1996). In consequence, the T cells transmit positive signals to the antigen-presenting centrocytes. Help provided by the $CD4^+$ germinal centre T cells, either in the form of direct cell surface interactions or in the form of cytokines, allows the centrocytes to survive and to switch the expression of their surface Ig from IgM to IgG or to further Igs by somatic recombination of Ig gene segments. The direct interactions of T and B lymphocytes in the germinal centre may involve the co-stimulatory CD28 molecule on T cells and its ligands CD80 and CD86 on B cells, as well as the CD40 ligand and CD40, respectively (Ferguson et al. 1996). Mice deficient for one of these receptor-ligand pairs are unable to form germinal centres and to mount IgG responses. It is, however, not clear up to now where the crucial interactions occur. Anyhow, the ability of centrocytes to activate help from their cognate $CD4^+$ T cells determines their survival or death. A selection process is initiated which finally leads to "affinity maturation" of the specific Ig produced: as the amount of antigen on FDCs is decreasing in the course of an immune reaction, there is a competition of centrocytes for antigen. Only those centrocytes with the highest affinity surface Ig achieved by mutation will finally internalise enough antigen to be efficiently "helped" by T cells (Stockinger et al. 1996). Centrocytes not receiving enough positive signals from T cells are prone to die by apoptosis close to the dark zone. It is assumed that this process of successive elimination of low-affinity B cells is yet another means of avoiding autoreactivity.

Besides positive interactions with T cells, centrocytes also receive additional signals from FDCs which reach beyond the provision of antigen. Thus, interactions of integrin and Ig family molecules on the surfaces of FDCs and B cells are essential for centrocyte survival (Kosco-Vilbois et al. 1993; Tew et al. 1997). Furthermore, complement fragments sticking to the immune complexes bound on the FDCs may engage complement receptor 2 on B lymphocytes and provide another survival-promoting signal, which may be especially relevant for B memory cell formation (Tew et al. 1997; Fehr et al. 1998). In addition, positive interactions between FDCs and B cells may also increase the antigen-presenting capacity of the B cells and thus their efficiency of eliciting T cell help (Kosco-Vilbois et al. 1997; Tew et al. 1997). The sequence of events with respect to selection of B lymphocytes is not entirely clear. There may be three successive control points where antigen-specific B cells are selected for survival, comprising first extrafollicular priming, then interaction with FDCs and finally, rescue of appropriate B cells by $CD4^+$ germinal centre T lymphocytes (Lindhout et al. 1997).

The fate of positively selected centrocytes is variable. Their further behaviour is already determined within the germinal centre, but it is uncertain which signals are operative. The descendants of centrocytes leave the germinal centre and are distributed in the body. A certain portion of these cells settle down in appropriate connective tissues, especially in the bone marrow, to locally differentiate into antibody-secreting long-lived plasma cells. These cells still undergo further affinity maturation of their secreted Ig even when germinal centres have waned in the spleen or lymph nodes after a single application of antigen (Takahashi et al. 1998). Thus, germinal centres are essential for Ig hypermutation, but long-term affinity maturation of secreted antibodies may proceed at other locations as well. Other centrocytes develop into memory B cells, which have the capacity of accelerated differentiation into plasma cells upon restimulation (see below). An unknown number of the memory cells joins the pool of recirculating B cells travelling across primary follicles or across the corona of secondary follicles to spread specific immunity within the body. Finally, it has been speculated that centrocytes may somehow return to the dark zone of the germinal centre, revert to centroblasts and resume proliferation (MacLennan et al. 1997). This speculation fits the theory that B cells undergo several rounds of hypermutation and immunoglobulin affinity selection in germinal centres before turning into plasma cells. Not only centrocytes, but also recirculating memory B cells might re-enter germinal centres after a renewed encounter with antigen (Berek and Milstein 1988). Thus, a recirculating centrocyte (if this cell exists) or a memory B cell will pass several germinal centres during its differentiation. Detailed kinetic data demonstrating this process are, however, lacking up to now. An alternative mechanism has recently been proposed by Camacho et al. (1998). At a certain stage of a primary T cell-dependent antibody reaction, proliferating B cells with specific surface Ig occur both in the dark zone and in the light zone of germinal centres in the spleen of mice. This indicates that B cell proliferation in the dark zone and subsequent selection in the light zone do not proceed in a one-way direction, but are repetitive events without spatial separation, at least initially. Obviously, a dark zone and a light zone cannot always be distinguished in primary splenic germinal centre reactions. This observation is in accordance with our findings in rats (see page 11).

The factors which program a centrocyte for development into a memory B cell are not entirely known (MacLennan et al. 1997). One prerequisite for memory cell formation is the interaction of the B cell CD40 surface receptor with the CD40 ligand on germinal centre T helper lymphocytes (Casamayor-Palleja et al. 1996; MacLennan et al. 1997). It is, however, clear that several additional stimuli coming from helper T cells, such as special combinations of cytokines, are required. That there is an intricate interplay among B and T lymphocytes within germinal centres is also demonstrated by the fact that undisturbed expression of CD86 (B7–2) on B cells contributes an essential prerequisite for B memory cell formation and Ig hypermutation in mice (Han et al. 1995). In addition, the interaction of CD40 ligand with B cell CD40, in combination with binding of appropriate cytokines, appears to program germinal centre B cells for memory cell differentiation in vitro, while in the absence of CD40 or its ligand the same cells will terminally differentiate into plasma cells (Liu and Banchereau 1997; Arpin et al. 1995). This effect is investigated in more detail by Randall et al. (1998) who demonstrate that ligation of CD40 on the B cell surface promotes in vitro activation, but arrests further differentiation to Ig-secreting plasma cells by selectively suppressing secretory Ig gene transcription. The authors interpret this phenomenon as a way to avoid the secretion of very large amounts of Ig within a germinal centre, which might mask the antigen on FDCs. It is, however, not clear whether masking of antigen represents a problem in vivo, because FDCs have some enigmatic capability of shielding their bound antigen from occupation by free antibody (Tew et al. 1997). Nevertheless, prevention of Ig secretion within germinal centres may be a desirable effect.

If the in vitro findings described for CD40 and its ligand also apply in vivo, the consensus model of B cell differentiation reviewed in this chapter suffers from some uncertainty about the term "plasma cell" . Thus, different investigators are rather vague about whether the extrafollicular foci and germinal centres contain plasma cells or not. Obviously cells with substantial amounts of specific intracellular antibodies are found at both sites. For germinal centres of the spleen this has been shown after repeated and prolonged immunisation in rabbits (Van Rooijen et al. 1984; Van Rooijen 1990). It has also been claimed that germinal centre cells secrete antibodies (Szakal et al. 1989). Thus, plasma cells may not regularly appear within germinal centres, but only under certain conditions, when memory B cells are pre-existing and antigen is present in excess. However, it has never been shown in detail that the cells in germinal centres really conform to the phenotype of mature plasma cells secreting immunoglobulin. In our opinion, it is decisive to distinguish "B cell blasts" (surface Ig-positive dividing cells) from "plasmablasts" (dividing cells with large amounts of intracellular Ig) and plasma cells (nondividing sessile end-stage secretory cells with intracellular Ig, but lacking surface Ig) as suggested by Liu et al. (1991). According to these criteria, which can, however, not be applied in immunohistology for technical reasons, the cells in germinal centres may have to be classified as plasmablasts, i. e. cells of a more immature stage not secreting Ig, which are still proliferating and have a migratory potential. The matter may be difficult to settle, because plasma cell differentiation proceeds continually. Thus, Smith et al. (1996) describe cells isolated from the spleens of hapten/carrier-immunised mice and most probably belonging to the extrafollicular foci which carry surface Ig while simultaneously secreting antibodies and exhibiting several cell surface markers of plasma cells.

5.1.3.1
Summary

The third step in the development of an antibody response in the spleen leads to the formation of germinal centres. This step requires circulating antigen-specific Ig, which is deposited in the form of immune complexes with antigen on the surface of FDCs in primary follicles. Recognition of the retained antigen by recirculating primed B lymphocytes leads to arrest and proliferation of these cells in the follicle and to the immigration of antigen-specific CD4$^+$ T helper cells. After some time, the proliferating B cells form the dark zone of the germinal centre and the cells are then called centroblasts. Centroblasts undergo a hypermutation process of the antigen-binding portion of their Ig which forms the basis for the selection of B lymphocytes with increasing affinity for the antigen. Before and after hypermutation, centroblasts are prone to apoptosis. Those centroblasts which survive further differentiate into centrocytes and pass into the light zone of the germinal centre. Centrocytes are positively and negatively selected for the affinity of their mutated surface Ig for the antigen and they are able to switch their Ig expression from IgM to other Igs. Both processes are accomplished by interaction with FDCs and CD4$^+$ T helper cells. Centrocytes are programmed to enter the B memory cell or the plasma cell pathway at some step of their differentiation before they leave the germinal centre. The signals determining this decision are under investigation.

5.1.4
Localisation of Memory B Cells in the Marginal Zone

Memory B cells have a lower threshold for activation and differentiation to plasma cells and a longer life-span than other B cells. The antigen-binding hypervariable regions of their surface Ig exhibit an increased affinity for antigen due to accumulated mutations in comparison to germ-line Ig. All memory cells are supposed to have passed germinal centres because the hypermutation process of the genes encoding the Ig antigen-binding regions only occurs at these sites. Unless stimulated by antigen, memory cells do not divide (Schittek and Rajewsky 1990). They are known to occupy defined strategically located regions of secondary lymphoid organs (Arpin et al. 1997; MacLennan et al. 1997; Spencer et al. 1998). Memory cells are able to recirculate, however, this property may only be relevant during certain stages of their life (Bachmann et al. 1994).

A special population of memory cells does not switch Ig class and continues expressing surface IgM. These cells or their precursors are able to recirculate and to colonise the splenic MZ and other secondary lymphoid organs. Obviously there is an advantage that MZ memory B cells (and also some long-lived plasma cells) continue expressing IgM and do not switch to production of other Igs. IgM is more protective than other Ig isotypes in immediate responses to bacteria due to its polyvalency and good complement-activating function. IgM-positive MZ B cells have undergone hypermutation and affinity maturation in germinal centres. This has not only been established in rodents, but also in humans. In man the hypervariable regions of IgM from B cells microdissected from the MZ contain accumulated mutations not found in the germ-line (Dunn-Walters et al.1995). Thus, in humans, mice and rats, the hypermutation mechanism obviously operates before Ig classes are switched during long-term antibody responses (Insel et al. 1994; Berek 1992). There may be a special „IgM-only“ pathway of memory cell differentiation in man. Hypermutated B cells of this phenotype form about 10% of the circulating B cells in human blood (Klein et al. 1997).

In contrast to germinal centres, which represent oligoclonal accumulations of B cells, the memory B cells in the splenic MZ in rats are derived polyclonally (Zhang et al. 1988). Thus, the MZ is composed of memory B cells of different specificities, including B cells specific for various T cell-dependent protein antigens as well as for T cell-independent polysaccharide antigens. In addition, "natural" i.e. pre-existing, autoreactive and xenoreactive antibodies against polysaccharide antigens may be produced by MZ B cells (Candinas et al. 1996; Chen et al. 1997). A substantial amount of the B cells producing IgM xenoreactive anti-polysaccharide antibodies are present in the spleen of rats. These antibodies are the major cause of hyperacute rejection in organ xenotransplantation. Interestingly, splenectomy substantially reduces circulating xenoreactive IgM and can ameliorate, but not abolish, hyperacute xenograft rejection in rats (Soares et al. 1995). During secondary T cell-dependent and T cell-independent antibody reactions in the spleen of mice and rats, memory B cells specific for the applied antigen leave the MZ (Liu et al. 1988). These cells transiently localise in the follicles and in extrafollicular foci of the oPALS before moving to the red pulp and developing into plasma cells (Claassen et al. 1986; Liu et al. 1991).

In rats, specific memory B cells are found in the MZ for long periods, i.e. for more than a year after antigen administration (Liu et al. 1991). Thus, under steady state conditions the majority of MZ memory cells is sessile and non-dividing after having arrived in their compartment (Gray et al. 1982; Liu et al. 1988). If memory B cells are removed from the MZ in rats, these cells are quickly replenished from recirculating precursors (Kumararatne and MacLennan 1981; Lane et al. 1986). Obviously, a certain proportion of IgM^+ memory B cells always recirculates between the secondary lymphoid organs and is able to colonise the MZ within 24 h of their last cell division (Chan and MacLennan 1993). In rats short-term ablation of MZ B lymphocytes can be achieved by cyclophosphamide or non-lethal irradiation (Kumararatne et al. 1980; Gray et al. 1984a; Lane et al. 1986). Injection of gram-negative bacteria or endotoxin leads to a quick depletion of MZ memory B cells within 4–6 h and restoration of the MZ within 24 h (Pettersen et al. 1967; Gray et al. 1984b).

The apparent contradictions concerning the kinetics of MZ memory B cells may be caused by the fact that the cells behave differently in the experimental situations mentioned above compared to the normal state. Rapid large-scale emigration and further differentiation of polyclonal MZ B cells makes sense during "emergency" reactions that are induced by contact with bacterial components such as lipopolysaccharide (Gray et al. 1984b; Groeneveld et al. 1985) or high circulating levels of cytokines, for example, provoked by interferon-gamma (Steiniger and van der Meide 1993). Obviously irradiation and cyclophosphamide also effect some kind of emergency signal for emigration, because the depletion of the allegedly sessile and non-dividing memory B cells from the MZ by these agents cannot be accounted for otherwise.

It is not clear, whether MZ memory B cells directly move to the oPALS or to the red pulp to differentiate into plasmablasts with certain types of stimulation or whether they have to first pass germinal centres. This may depend on the presence of pre-existing germinal centres. Whether MZ B cells reach the follicles by direct immigration, via the blood, or by both routes, is also not known. Interestingly, in humans, B memory cells can directly differentiate into plasma cells in vitro if adequate cytokines and an initial stimulus via CD40 are given (Arpin et al. 1997). This shows that CD40 ligation provokes different effects in memory B cells as compared to germinal centre B cells (i.e. centrocytes), which are prevented from plasma cell differentiation after the interaction of CD40 with its ligand.

There is a long-lasting and continuing controversy as to whether persisting antigen and help by T cells is required for the maintenance of B cell memory (Gray and Skarvall 1988, Schittek et al. 1990, Ahmed and Gray 1996). It is well-known that specific serum antibodies may persist throughout the lifespan of rodents after appropriate immunisation (Ahmed and Gray 1996). Either of three mechanisms may account for this phenomenon: memory B cells or their progeny may be long-lived; long-term antigen reservoirs may exist in the body for continuous restimulation of memory cells; or there are plasma cells with extremely long life-spans. Obviously, much additional research is needed to understand which of these possibilities is relevant.

5.1.4.1 Summary

Among other possibilities a centrocyte may decide to further develop into a memory cell expressing IgM, but not IgD. Such $IgM^{+}IgD^{-}$ B memory cells are the result of a special differentiation pathway present in rodents and in man which leads to the preferential accumulation of the cells in the MZ of the spleen. The cells are, however, also present in the blood and in other secondary lymphatic organs. MZ memory B cells are sessile non-dividing cells in the steady state. Nevertheless, they can be easily induced by bacterial products and by certain cytokines to emigrate from the MZ and to differentiate into plasma cells. Subsequently, the MZ is quickly reconstituted by recirculating IgM^{+} memory B cell precursors.

5.2 Function of the Marginal Zone during Primary Antibody Responses against T Cell-Independent Type 2 Antigens

In humans the presence of the spleen is often essential for primary antibody responses against encapsulated bacteria (Hosea et al.1981; Timens 1991). After splenectomy, formation of such antibodies is not always totally abrogated, but the amount of Ig produced is diminished. This fact leads to an increased risk of a fulminant septic syndrome in asplenic patients, especially in children (Shaw and Print 1989). The risk is further augmented by the absence of a large organ responsible for phagocytosis of opsonised bacteria.

The response against encapsulated micro-organisms is primarily directed against neutral polysaccharides of their capsules, which have been classified as T cell-independent type 2 (TI-2) antigens (Hosea et al. 1981). TI-2 antigens consist of a backbone of polysaccharides carrying a large number of repetitive epitopes, i.e. binding sites for Ig (Humphrey 1985). It is widely accepted that MZ B cells are the most important responsive cell population. In general, the antibody response against TI antigens is assumed to proceed without formation of germinal centres or memory cells. However, Zhang et al. (1988) and Liu et al. (1991) have shown that TI-2 antigens provoke specific memory cells in the MZ of rats when immunisation proceeds in the presence of other bacterial components such as lipopolysaccharide and that a somewhat reduced formation of germinal centres is also involved. This is in agreement with the fact that the requirement for T cell help during antibody production against TI-2 antigens is reduced in comparison to T-dependent antigens, but it is not totally lacking (Mond et al. 1995; Snapper and Mond 1996). Obviously, cytokines or other signals from non-T cells are most decisive for promoting antibody production against TI-2 antigens. Thus, it is still open to speculation, whether the B cell population responding to TI-2 antigens represents memory cells primed by environmental bacterial antigens or a totally separate B cell lineage in the MZ (Spencer et al. 1998). The response to TI-2 antigens is absent in the CBA/N mouse strain carrying an X chromosome-linked immune defect caused by mutation of a tyrosine kinase (Mosier and Subbarao 1982). TI-2 antigens do not necessitate presentation to T cells by MHC class II molecules (Mond et al. 1995). This special behaviour of TI-2 antigens may be due to the high epitope number, to the special properties of the MZ B cells stimulated, or to a special interaction of B cells and macrophages or other cells in this region (Liu 1997).

How TI-2 antigens stimulate MZ B cells is still controversial. In rats neutral TI-2 polysaccharide antigens are selectively taken up by MZMϕ and by MMMϕ, whereas in mice only MZMϕ appear to ingest these antigens (Humphrey and Grennan 1981; Matsuno 1989). In mice, labelled TI-2 antigens lodge in MZMϕ for prolonged times after intraperitoneal injection and it has been speculated that these cells are essential for stimulation of specific MZ B cells (Humphrey 1981; Humphrey and Grennan 1981). It was also shown in mice that the uptake of certain TI-2 antigens in MZMϕ is blocked in vitro by the polysaccharides mannan and zymosan (Chao and MacPherson 1990). However, the role of MZMϕ and MMMϕ in TI-2 antibody responses is far from clear. In mice MMMϕ can be distinguished from MZMϕ by specific MAbs (Van Vliet et al. 1985; Dijkstra et al. 1985b; Kraal and Janse 1986). Attempts to abolish either MZMϕ or MMMϕ by in vivo antibody treatment or by other methods have led to controversial results questioning the requirement of MZMϕ for antibody responses against TI-2 antigens (Kraal et al. 1988; Kraal et al. 1989a; Buiting et al. 1997). Those studies showing a crucial role for MZMϕ are prone to criticism because systemic treatment with anti-macrophage antibodies or other macrophage-depleting agents may lead to consequences that reach far beyond the spleen. Removal of macrophages may also affect the minor contribution of helper T cells to TI-2 antibody reactions by impairing macrophage antigen presentation. It is open to speculation which cell-surface receptors on MZMϕ play a role for capture of TI-2 antigens.

In rats, MZMϕ and MMMϕ are phenotypically indistinguishable up to now and both cell types are characterised by strong expression of sialoadhesin (Kraal 1992). Red pulp macrophages, on the contrary, only carry minor amounts of this molecule. This is similar to the distribution found in mouse spleens (Crocker and Gordon 1989). In mice, sialoadhesin may be somehow involved in the adhesion of bacteria to MZMϕ mediated by soluble fragments of the macrophage mannose receptor (Martinez-Pomares et al. 1996). Most investigators, however, view the role of sialoadhesin in the context of lymphocyte migration. In rats, adhesion assays on spleen sections show that sialoadhesin mediates interaction of lymphocytes especially with MMMϕ (van den Berg et al. 1992). Whether this is also important in vivo has so far not been assessed.

Mouse MZMϕ have even been compared to the endothelium of high endothelial venules in lymph nodes with respect to lymphocyte recirculation (Lyons and Parish 1995).

Whether macrophages in the MZ are of any importance for antibody production against TI-2 antigens has also been questioned for other reasons. TI-2 antigens bind complement and may stick to MZ B cells via B cell complement receptors (van den Eertwegh et al. 1992b). Binding of TI-2 antigens to MZ B cells with the help of complement also occurs in humans (Peset Llopis et al. 1996). Thus, MZ B cells may be directly activated by TI-2 antigens via immune complexes and/or complement receptors. The function of MMMϕ and MZMϕ in mice and rats still remains to be established.

5.2.1 Summary

Splenectomy causes a defect in antibody responses to T-independent type 2 polysaccharide antigens, which occur in bacterial capsules. It is controversial whether certain B memory cells in the MZ are responsible for rapid antibody production against TI-2 antigens or whether the B cells reactive with TI-2 antigens represent a totally separate population. In rodents, TI-2 antigens localise in MZMϕ after systemic application, but the antigens can also directly bind to MZ B cells in vitro. The functional role of MZMϕ and MMMϕ in general and for TI-2 antibody responses in particular has not been unequivocally established up to now.

5.3 Function of the Red Pulp

In rats and in man, the splenic red pulp consists of sinuses and splenic cords. Blood arrives in the sinuses either directly from connections with the finest branches of the arterial tree or indirectly from the splenic cords. The former pathway is regarded as the "closed" part of the splenic circulation while the latter forms the "open" part. The open splenic circulation of rats consists of arterioles ending in the splenic cords (Satodate et al. 1986) and blood flowing through the cords to enter the sinuses from the outside via interendothelial slits in the sinus walls (Chen and Weiss 1973; MacDonald et al. 1987). The same appears to be true for man (Weiss et al. 1985).

The splenic cords are composed of reticular fibres covered by fibroblasts, and of thrombocytes, sessile macrophages and other leukocytes (Kashimura and Fujita 1987). It is supposed that the filtration function of the spleen with respect to larger particles and aged or altered red cells primarily resides in the meshwork of the cords and, perhaps, also in the walls of the sinuses. The macrophages in the cords are regarded as the cells being responsible for uptake and removal of debris and altered cells filtered from the blood. Abnormalities of red cells such as spherocytosis, infection with malaria parasites or antibody deposition on their membranes lead to an abnormal erythrocyte sequestration and destruction in the red pulp provoking hypersplenism and anaemia. Similar phenomena may also occur with respect to platelets. These pathologies may be corrected by splenectomy.

In humans and probably also in rats, the spleen does not serve as a storage organ for erythrocytes as it does in some mammals, which attain high running velocities for short periods or are adapted to diving. In these species, such as horses, cats, dogs, deer and seals, contraction of the spleen upon adrenergic stimulation results in an "autotransfusion" of erythrocytes which improves oxygen supply to the tissues. In humans, the only blood component normally stored in the spleen are thrombocytes. About one third of the exchangeable thrombocyte pool is normally present in the human spleen and can be mobilised if necessary (Jandl and Aster 1967; Peters et al. 1980; Du P. Heyns et al. 1985; Kotzé et al. 1986). Whether normal granulocytes are also pooled, is not clear. It has been estimated that about 50% of the human marginal pool of granulocytes is residing in the spleen with a mean transit time of about 15 min (Peters et al. 1985). There is no consensus as to whether the human splenic red pulp is a normal storage site for leukocytes which may be released during exercise (Baum et al. 1996; Nielsen et al. 1997). As mentioned above, pooling of mature erythrocytes in normal rodent and human spleens is controversial (see Sect. 8.2). Most investigators agree that reticulocytes are, however, retained in the human spleen for final maturation.

Neither in rats nor in humans is there any information on the kinetics or function of the large number of B lymphocytes and T lymphocytes which occur in the splenic cords. It is not clear whether these cells are just migrating through the cords in the process of leaving the spleen or whether some of these cells are sessile. In rats and in man, B lymphocytes in the red pulp preferentially occur in "sheaths" surrounding arterioles. Thus, there seems to be a normal B cell compartment in the red pulp which has so far been ignored in most investigations. It is unknown which part of the arterial tree is covered by this sheath. In experimental animals the surroundings of terminal arterioles in the red pulp harbour the majority of antibody-containing cells after secondary immunisation (van Rooijen et al. 1984). However, most authors do not exactly define where these terminal arterioles originate or end. It has always been supposed that the B cells in the vicinity of arterioles are antibody-containing stimulated B cells migrating to the red pulp cords to become plasma cells. Nieuwenhuis and Ford (1976) mention that a few labelled thoracic duct B lymphocytes recirculating through the rat spleen also distribute to the surroundings of "penicillar arterioles" at 18 h after i.v. injection and later. To our knowledge, it has never been investigated in more detail whether small red pulp arterioles provide a compartment for slow migration of recirculating B lymphocytes.

The most important organ for long-term production of high-affinity serum antibodies is the bone marrow (Takahashi et al. 1998). The contribution of the splenic red pulp to long-term antibody production seems to be only minor. The plasma cells in the splenic red pulp are still rather ill-defined with respect to their origin and life-span. Ho et al. (1986) found that 75% of the cells containing cytoplasmic Ig in the red pulp of normal non-immunised rats were labelled after continuous infusion of [^{3}H]thymidine for 3 days. The labelling index increased to 90% after treatment for 10 days and contrasted with the labelling of only 10% and 40%, respectively, of Ig-positive cells in the bone marrow. These results were interpreted as indicating that the half-life of the majority of plasma cells in the splenic red pulp is less than 3 days, while it is more than 3 weeks in the bone marrow. This opinion fits the current model of B cell responses in the spleen, but the investigation of Ho et al. (1986) may underestimate the life-span of at least some proportion of the red pulp plasma cells. The cumulative incorporation of nuclear label does not necessarily indicate that all labelled cells will die quickly. Thus, unless the long-term loss of the label is assessed, it cannot be excluded that there is a plasma cell population which enters a long-lived non-replicating state after initial division.

5.3.1 Summary

In rats and in man, the splenic red pulp consists of cords and sinuses. The cords represent the open part of the circulation system in the spleen. Among other cell types, they harbour the majority of the splenic macrophages. The cords and the perforated sinus walls are supposed to be the sites of clearance of altered erythrocytes and platelets. Plasma cells occur exclusively in the cords. The life-span of red pulp plasma cells has been estimated to be on the order of 3 days in rats. There is a substantial B cell compartment around small arterioles in the splenic red pulp. In humans, the splenic red pulp is a normal site of thrombocyte pooling, but most investigators deny that this is also true for granulocytes or erythrocytes.

5.4 Role of the Spleen in CD8$^+$ Cytotoxic T Cell Responses

In contrast to B lymphocytes, interest in the behaviour or distribution of antigen-specific CD4$^+$ or CD8$^+$ T cells in the spleen has been rather low. This is primarily due to the problem that MHC restriction precludes the direct visualisation of antigen-specific TCRs by application of the antigen to tissue sections. There are MAbs recognising families of alpha or beta chain TCR variable regions, which are preferentially used by T cells for binding certain antigen-MHC complexes, but these reagents do not unequivocally demonstrate the specificity of individual T cells. Loss of the spleen does not obviously affect cytotoxic T cell immunity. The prerequisites for induction of cytotoxic T cell responses, such as dendritic cells bearing high amounts of MHC class I antigens, are present in the spleen. Besides presenting antigens to CD4$^+$ T cells in the context of MHC class II molecules, dendritic cells are also very potent stimulators of naive MHC class I-restricted CD8$^+$ cytotoxic T lymphocytes (Porgador and Gilboa 1995; Mitchell et al. 1998).

A contribution of the spleen to cytotoxic T cell responses may be present, but it is difficult to demonstrate. The spleen filters out large amounts of viral antigens which can lead to activation of unprimed cytotoxic T cells in the organ within 24 h. The spleen also represents a major reservoir of memory T cells for cytotoxic reactions against viruses or viral-encoded proteins in mice (Kündig et al. 1996a). Thus, the organ provides an environment for the accumulation of antigen, for rapid priming of unstimulated cytotoxic T cells and also for restimulation of cytotoxic memory T cells, if viral antigens arrive via the blood. Such an immediate reaction is obviously not possible in lymph nodes or other peripheral organs invaded by viruses. Protective cytotoxic T cell immunity only occurs at the latter sites when major numbers of activated T cells have arrived by recirculation and emigration via high endothelial venules or other vessels (Kündig et al. 1996b). If cytotoxic memory T cells depend on the amount of antigen arriving and perhaps also persisting in a given lymphatic organ, the spleen appears to be of crucial importance for long-term maintenance of antiviral protection by cytotoxic T cells in vivo, at least in mice. This splenic function is, however, not directly apparent during systemic viral infections, because the most important initial antiviral protection is achieved by circulating antibodies. Obviously, special experimental settings avoiding recognition of viral epitopes by antibodies and utilising a functional in vivo assay have to be applied for demonstrating an effect of the spleen on cytotoxic T cell immunity. It would be interesting to know whether loss of the spleen leads to defective memory T cell reactions against certain viruses. Interestingly, Kündig et al. (1996a) have shown that mice possessing antiviral cytotoxic memory T cells in the spleen are protected against systemic viral re-infection, but not against viral replication in peripheral lymph nodes and other peripheral organs as long as the virus stays in these sites. This indicates that either antigen retention or the immigration pathway of recirculating cytotoxic memory T cells differ among the spleen and, for example, peripheral lymph nodes.

5.4.1 Summary

The role of the spleen in cytotoxic T cell reactions is difficult to assess. Little information is available, because special experimental systems are needed. Preliminary results obtained with antiviral cytotoxic T cells in mice indicate that the spleen is a major site of deposition of viral antigens and that antiviral cytotoxic memory T cells primarily reside in this organ.

5.5 The Spleen, Natural Killer Cells and Gamma/Delta T Cells

Most reports dealing with the distribution of natural killer (NK) cells in humans agree on the fact that the spleen can be a reservoir for cells capable of exerting natural cytotoxicity. NK cells form a lymphocyte population distinct from $CD4^+$ or $CD8^+$ T cells and B cells, which is able to lyse target cells without priming and in the absence of MHC restriction. In fact, NK cells are often negatively regulated by the recognition of MHC antigens on target cells. The phenotypic definition of NK cells is still somewhat problematic, because there are overlaps with a special population of $CD8^+$ alpha/beta TCR^+ T cells (so-called "natural T cells" , Bix and Locksley 1995). In addition, some natural cytotoxic functions such as tumour cell lysis, may also be exerted by other cell types including macrophages. There has been a tremendous increase in knowledge about the functions and receptors of NK cells during recent years (Lanier 1998). It is still not entirely clear, how many subpopulations of NK cells exist in man. The majority of cells exhibiting an "NK cell" phenotype in the human spleen appears to reside in the red pulp with a few cells also present in the follicles and germinal centres (Witte et al. 1989). However, similar to T cells, there are no detailed immunohistological investigations on the intrasplenic distribution of different NK cell populations.

Interestingly, T lymphocytes exhibiting the gamma/delta TCR also preferentially occur in the splenic red pulp, but not in other compartments (Bordessoule et al. 1990).

5.5.1
Summary

NK cells and T cells carrying the gamma/delta T cell receptor preferentially reside in the red pulp of the human spleen.

6 Recirculation of Lymphocytes Through the Spleen

The spleen is the most important organ of lymphocyte recirculation in man and rodents (Pabst 1988a). Recirculation implies the distribution of lymphocytes to all secondary lymphatic organs and other tissues by the blood, their emigration from the vasculature, a sojourn in the respective tissue and the subsequent return into the blood. In most tissues the route of lymphocyte return to the blood involves the lymphatic vessels and finally, the thoracic duct or other large lymph ducts joining the venous system. The case of the spleen is somewhat different, because recirculation implies extravasation, a period of lymphocyte arrest in the white pulp and subsequent direct re-entry into the venous effluent from the organ. Larger efferent lymphatics are clearly present in the vicinity of large arteries in the spleen of man and rodents (own observation; Pellas and Weiss 1990) and lymph capillaries occur in the human PALS (Van Krieken et al. 1985), but it is assumed that lymphatic drainage only plays a negligible role for lymphocyte recirculation through the spleen.

Recirculation is essential to allow lymphocytes to survey sites of antigen presentation, to permit optimal cooperation among T and B cells, to keep the recirculating cells in a state of immunological competence and to permit spreading of immune reactions. It has been estimated that in humans the number of lymphocytes migrating through the spleen exceeds that of cells recirculating through all lymph nodes by a factor of 10 to 20 (Pabst 1988b). The mean transit time of lymphocytes through the blood is about 25 min in man and there is a 48-fold exchange of the blood lymphocyte pool each day. It has been estimated that 10%–15% of the exchangeable B cell pool and about 25% of the T cell pool resides in the human spleen (Egelund Christensen et al. 1978).

Lymphocyte kinetics and migration pathways within the spleen are totally unknown in man, but lymphocyte entry into the spleen has been investigated in some detail in rodents. In rats, the accumulation of intravenously injected labelled thoracic duct lymphocytes in the spleen is about 10-fold faster than in lymph nodes (Smith and Ford 1983). In rat and mouse spleens, the MZ is regarded as the entry compartment of the white pulp, because labelled B and T lymphocytes rapidly appear in the MZ within 15 min after intravenous injection (Van Ewijk and Nieuwenhuis 1985; Nieuwenhuis and Ford 1976; Brelinska and Pilgrim 1983). In principle, there are two ways by which lymphocytes may arrive in the MZ: they may leave the marginal sinus through its rather leaky MZ side or they may be poured directly into the MZ by capillaries ending there (Veerman and van Ewijk 1975). Thus, it is not clear whether adhesion to endothelial cells is needed at all to arrest recirculating lymphocytes in the MZ. In mice, it has been shown that a special endothelial adhesion glycoprotein, MAdCAM-1, which is of crucial importance for mucosal „homing“ of lymphocytes, is expressed on the inner endothelium and additional cells of the marginal sinus (Kraal et al. 1995). However, a function of this molecule for lymphocyte immigration into the splenic white pulp was not found. Thus, it cannot be excluded that instead of endothelia, fibroblasts or their products, such as certain chemokines, mediate lymphocyte arrest in the MZ of rodents and direct the migration of these cells.

It is well established that recirculating B cells move from the MZ into the oPALS and from there into the corona of secondary follicles or into primary follicles. In rats, this journey may take more than 18 h after intravenous injection of labelled lymphocytes (Nieuwenhuis and Ford 1976; Van Ewijk and Nieuwenhuis 1985; Willführ et al. 1990). T cells also arrive in the MZ, but they quickly head for the PALS where they can be found from 1 h after injection onwards with a slow decline in numbers beginning after about 6 h (Goldschneider and McGregor 1968; Nieuwenhuis and Ford 1976; Willführ et al. 1990). Finally, in rats both B and T cells may move along so-called terminal arterioles or MZ bridging channels, which represent side branches of the central arterioles across the MZ, to reach the red pulp and leave the organ (Mitchell 1973; Liu and Banchereau 1996). The exit route from the white pulp via the MZ bridging channels has not been investigated in detail up to now and it is primarily deduced from immunohistology.

The transit time of intravenously injected labelled thoracic duct T lymphocytes through the spleen is about 4–5 h in rats (Smith and Ford 1983). As mentioned above, the transit takes longer for B cells. It has to be kept in mind that the details of this process are the following: after a short interval, only a minority of about 10% of the leukocytes arriving by the blood immigrate into the white pulp in rats, while 90% of the cells end up in the red pulp (Kraal et al. 1989b). This 90% has a short transit time of about 5 min, i.e. they are quickly washed out, probably via the sinusoids. Nevertheless, as 3%–5% of the cardiac output goes to the spleen in man (Llende et al. 1986, Pabst 1988b), repeated passages through the organ lead to a substantial extraction of lymphocytes. Interestingly, when labelled thoracic duct lymphocytes are injected intravenously in rats, there is always a high number of cells lodging in the red pulp. Thus, the majority of transferred B cells is always found in the red pulp during the initial 24 h (Willführ et al.1990). It has also been reported that the distribution of $CD4^+$ and $CD8^+$ T cells in the rat spleen is not identical within 24 h after injection. More $CD4^+$ cells than $CD8^+$ cells accumulate in the PALS, while $CD8^+$ cells have a preference for the MZ and red pulp (Willführ et al. 1990). In addition, a small proportion of $CD4^+$ and $CD8^+$ thoracic duct lymphocytes always migrate towards the follicles (Willführ et al. 1990).

Thus, lymphocyte entry and distribution in the white pulp are rather well-known in rats. It is, however, astonishing that there is almost no information on the behaviour of recirculating lymphocytes in the splenic red pulp, which harbours a large population of these cells. In addition, detailed information as to how lymphocytes leave the white pulp again is also lacking. This may be a relevant topic to study because the spreading of immunological information by recirculation of activated or memory lymphocytes may not only be controlled at the point of entry but also at the exit pathway from a lymphatic organ.

6.1 Summary

In rats and mice, recirculating B and T lymphocytes enter the spleen via the marginal sinus. Labelled lymphocytes injected intravenously arrive in the MZ within 15 min after transfer. B lymphocytes then migrate into the oPALS and from there into primary follicles or into the corona of secondary follicles. This journey takes more than 18 h in rats. T lymphocytes immigrate into the PALS from the MZ within 5–6 h after injection. Both types of lymphocyte leave the spleen via the red pulp. Although the spleen is the major organ of lymphocyte recirculation, no information about migration pathways in human spleens is available.

7 The Role of Cytokines and Chemokines in the Development of Splenic Compartments

For the past 2 years, the mouse spleen has been the organ most thoroughly investigated to elucidate the role of cytokines and chemokines in influencing the structure of secondary lymphoid organs during ontogeny and in the adult state. Mice with deficiencies of cytokine genes and animals injected with soluble cytokine receptors before birth have provided the first evidence that the normal development of splenic compartments depends on the action of diffusible (or membrane-bound) mediators. At present, there are two types of molecules, cytokines of the TNF family and several chemokines, which have been shown to influence the compartmentalisation of the spleen in mice.

In mice, cytokines of the TNF family and their receptors are involved in mediating the communication of sessile and mobile cells during the formation of lymphoid organs at certain crucial steps in ontogeny. The TNF family of cytokines sensu stricto comprises three different proteins, TNF-alpha, lymphotoxin-alpha (formerly called TNF-beta) and lymphotoxin-beta (Ware et al. 1995). The molecules have pleiotropic immunostimulatory and growth-promoting effects, but they can also exert cytotoxicity on receptor-bearing target cells. The effects of members of the TNF family differ according to the type and the activation and/or differentiation stage of the receptor-bearing cell.

Bioactive TNF-alpha occurs both as a transmembrane molecule and as a soluble homotrimer, which assembles after proteolytic cleavage of the membrane form by metalloproteases (Gearing et al. 1994). In mice and man, both forms of TNF-alpha bind to two types of membrane receptors, the type 1 (p55 and p60, respectively) or the type 2 (p75 and p80, respectively) TNF receptor (Grell et al. 1995). Lymphotoxin-alpha is assumed to primarily exist as a homotrimer in solution. It binds to the same two receptors as TNF-alpha. However, lymphotoxin-alpha can also be found on cell membranes. In this case, one molecule of lymphotoxin-alpha is bound to two molecules of lymphotoxin-beta during biosynthesis, so that lymphotoxin-beta represents the membrane-spanning component of the complex. In addition, there is a minor form of two molecules of lymphotoxin-alpha associated with one molecule of lymphotoxin-beta. The lymphotoxin-alpha/beta complex binds to a third type of receptor called the lymphotoxin-beta receptor (Crowe et al.1994; Browning et al. 1995).

When the gene for TNF-alpha is deleted in mice, the animals have a normal T cell compartment in the spleen, but they totally lack primary or secondary follicles (Pasparakis et al. 1996). Networks of FDCs are absent. The mice are able to switch their antibody responses to IgG, but they exhibit very much impaired IgG responses to T cell-dependent antigens. However, the MZ is expanded and MMMϕ are present.

Interestingly, the IgG response to TI-2 antigens is enhanced. The structure of lymph nodes is grossly normal.

Similar findings occur in mice with a deletion of the TNF-receptor type 1 (p55). This TNF-receptor defect cannot be corrected by transfer of wild-type bone marrow cells (Matsumoto et al. 1997a) and it additionally leads to an altered distribution of plasma cells (Tkachuk et al. 1998). In contrast, deletion of the TNF-receptor type 2 (p75), does not abolish FDCs or the presence of follicles (Le Hir et al. 1996).

Interestingly, deletion of the gene for lymphotoxin-alpha has the most profound effect on splenic organisation and on the occurrence of other secondary lymphoid organs. The lack of lymphotoxin-alpha during ontogeny totally abolishes the ordered microanatomical structure of the spleen (Matsumoto et al. 1997b). There is no segregation of T and B cells into separate zones any longer. Primary or secondary follicles, FDCs and MMMϕ are also lacking. In addition, lymph nodes and Peyer's patches do not develop at all. Lymphotoxin-alpha-deficient mice are unable to switch their humoral immune responses to IgG. It is controversial, whether high antigen doses may overcome this defect (Matsumoto et al. 1996a). Reconstitution with wild-type bone marrow cells is, however, able to correct splenic microarchitecture in lymphotoxin-alpha-deficient mice (Matsumoto et al. 1997a, Tkachuk et al. 1998). It has recently been shown that in this case, the cells responsible for correcting the defect in formation of splenic follicles and FDCs and the defect in IgG production are lymphotoxin-alpha expressing B cells, but not T cells (Fu et al. 1998).

The deletion of the gene for lymphotoxin-beta leads to a lack of some, but not all, lymph nodes and to a disturbed arrangement of the B cell compartments of the spleen, while the T cell area is not much altered (Alimzhanov et al. 1997; Koni and Flavell 1998).

Genetic inactivation of the lymphotoxin-beta receptor apparently provokes a phenotype similar to that of lymphotoxin-alpha knockout mice, with a total abolition of separate T and B cell zones in the spleen (Pfeffer et al. 1997). However, if the function of the lymphotoxin-beta receptor is antagonised by prenatal treatment of mice with a soluble lymphotoxin-beta receptor, not all lymph nodes are abolished and lymph nodes draining mucosal surfaces still develop (Rennert et al. 1997).

Thus, cytokines of the TNF family influence the ontogenetic development of splenic compartments, lymph nodes and Peyer's patches in an organ-specific way. The experimental results available from mice deficient in cytokines or cytokine receptors of the TNF family are difficult to interpret at the present state of knowledge. It is obvious that the expression of lymphotoxin-alpha and TNF-alpha by B cells is essential for the proper structure of splenic secondary follicles and that lymphotoxin-alpha is needed for MMMϕ development. However, mice with genetic deletions of TNF family members and their receptors reveal a complex picture when the ontogeny of lymph nodes is also considered. Thus, for example, mesenteric lymph nodes are absent in lymphotoxin-alpha deficient mice, but they are present in lymphotoxin-beta deficient mice and in mice lacking TNF receptor 1 and TNF receptor 2. The effects on lymph node development can thus only be explained by either postulating so far unknown receptors for TNF family members, an unanticipated binding behaviour of the known TNF family cytokines or synergistic actions of their receptors (Koni and Flavell 1998).

Obviously, the facts mentioned represent pieces of a puzzle which still await a better characterisation of the TNF cytokine family and the relevant receptors before they can be assembled in a meaningful way. In addition, the causalities may be even more complicated, because it is not clear whether the effects of deficient genes are direct or indirect and at what stage of development they occur. Transfer experiments with wild-type lymphocytes or bone marrow indicate that there are no intrinsic maturation defects in stromal cells if cytokines of the TNF family are genetically absent (Matsumoto et al. 1996b).

This is also supported by another model, which is independent of effects during ontogeny. Adult severe combined immunodeficiency (SCID) mice lack mature T and B lymphocytes and do not exhibit phenotypically detectable FDCs. Transferred T and B lymphocytes from wild-type mice can, however, form relatively normal splenic T and B cell compartments in these animals. The transfer of wild-type B lymphocytes provokes the appearance of host-type FDCs if the B cells express lymphotoxin-alpha. In reconstituted SCID mice, endogenous TNF-alpha and lymphotoxin-alpha and -beta are present and these mediators permit the segregation of transferred T and B cells into their appropriate compartments in the spleen. This segregation is, however, inhibited when signalling via the lymphotoxin-beta receptor is blocked by application of soluble lymphotoxin-beta receptors (Gonzalez et al. 1998). In this case, phenotypically mature FDCs do not arise. Thus, also in the SCID model, lymphotoxin-alpha produced by B lymphocytes and the lymphotoxin-beta receptor on host stromal cells are a prerequisite for the normal formation of B and T lymphocyte compartments and development of FDCs in the spleen.

Alteration of the TNF family mediators leads to defective expression of adhesion molecules for lymphocyte emigration from the vasculature. It has been speculated that this fact causes the abnormal composition of lymphoid organs. Functional studies have, however, so far been unable to identify adhesion molecules or cells causally involved in abnormal lymphocyte segregation in knockout mice (Rennert et al. 1997). Thus, it is likely that TNF family gene deficiencies have indirect consequences such as disturbed production of B or T cell-specific chemotactic factors responsible for cellular segregation. The role of TNF family cytokines in the organisation of lymphatic organs will become more evident when gene deficiencies are introduced which can be induced in adult animals. This will circumvent possible adaptation of cellular interactions during the ontogeny of knockout mice.

Chemokines are a special type of cytokines. They represent low molecular weight protein mediators which provoke chemotactic migration and activation of leukocytes exposed to chemokine concentration gradients (Baggiolini et al. 1997). The receptors for chemokines are molecules with seven transmembrane domains and signalling pathways coupled to G-proteins. At present, four families of chemokines, named C, CC, CXC and CXXXC chemokines, are distinguished according to the number and spacing of cysteine residues in their NH_2-terminal portions (Mackay 1997). Chemokine gradients occur in soluble and also in more or less stationary forms because chemokines can bind to cell surface or extracellular matrix proteoglycans. A large number of chemokines is involved in leukocyte adhesion to endothelial cells and leukocyte activation during emigration from vessel walls (Rot 1993).

The first chemokine shown to have an exclusive action on lymphocytes, with a preference for B cells, is BLC (B lymphocyte chemoattractant) in the mouse or BCA-1 (B cell-attracting chemokine 1) in humans. BLC/BCA-1 is a CXC chemokine expressed by FDCs in splenic follicles, Peyer's patches and in some, but not all, lymph nodes (Gunn et al. 1998; Legler et al.1998). BLC/BCA-1 binds to a receptor on B cells called Burkitt's lymphoma receptor 1 (BLR1) or CXC receptor 5 (CXCR5). The mouse homologue of BLR1 regulates B lymphocyte entry into the follicles of the spleen and of Peyer's patches, but not of lymph nodes (Förster et al. 1996). Animals defective for BLR1 exhibit an altered composition of splenic compartments, and labelled B lymphocytes from BLR1 knockout mice are unable to enter splenic follicles in wild-type mice in vivo. The report by Förster et al. (1996) about BLR1 was the first indicating that the immigration of lymphocytes into splenic compartments may be induced by gradients of chemotactic mediators in a more ore less organ- and cell-specific way. The identification of lymphocyte-specific chemokines explains the total inhibition of B and T lymphocyte entry into the splenic white pulp after pre-treatment of transferred mouse lymphocytes with pertussis toxin in vitro (Cyster and Goodnow 1995; Lyons and Parish 1995). Pertussis toxin is known to inactivate certain G-proteins involved in chemokine signalling.

In addition to BLC/BCA-1, several new chemokines expressed constitutively by IDCs or FDCs have meanwhile been shown to attract T and B lymphocytes to secondary lymphoid organs and to promote migration to the PALS or to the follicles, respectively (Ngo et al. 1998; Schaniel et al. 1998). In vitro studies have identified the CXC chemokine SDF-1 (stromal cell-derived factor-1) as the first molecule inducing chemotaxis of recirculating B lymphocytes, but not of germinal centre B cells in humans (Bleul et al. 1998). This chemokine was shown to be expressed in certain stromal cells surrounding, for example, the germinal centres of human tonsils. Thus, the world of chemoattractant mediators initiating the specific distribution of different lymphocyte subsets to their microanatomical compartments in the spleen and in other lymphoid organs is just beginning to be explored. It may be that chemokine gradients are sufficient to establish the three-dimensional arrangement of immunocompetent cells in lymphatic organs. How the production of cell-specific chemokines is regulated in the steady state and during immune reactions, and which mutual cellular interactions are necessary to attract B cells to the follicles and T cells to the PALS will remain a difficult task to unravel.

7.1 Summary

During recent years the establishment of gene-deficient mice has shown that cytokines of the TNF family and their receptors and cell-specific chemokines play an essential role for the organisation of the spleen into compartments primarily populated by T or B lymphocytes. Thus, lymphotoxin-alpha, a member of the TNF cytokine family, which is also expressed by B cells, was demonstrated to be essential for the overall structure of the splenic white pulp and, in particular, for the development of splenic FDCs and follicles during ontogeny. Vice versa, a more or less B cell-specific chemokine was identified in mice and humans, which is produced by FDCs and attracts B cells into follicles. There are indications that certain stromal cells produce a chemokine which attracts only recirculating B cells into the follicle corona and that IDCs secrete chemokines with a selective action on activated T cells. At present it is clear that there must be several cell-type and/or organ-specific soluble mediators produced by migratory lymphocytes or sessile stromal cells, which are responsible for shaping the splenic white pulp. How many of these factors exist and in which sequence they act will be revealed in the future.

8 Unsolved Problems of Human Splenic Structure and Function

8.1 Arterial Blood Supply to the Splenic Follicles and to the Perifollicular Zone

The structure of the splenic arterial tree and the organisation of splenic blood circulation is still incompletely known in humans. In comparison to rats there is a much more extensive branching of the intrasplenic arteries and arterioles in humans. Our preliminary results indicate that, for example, human splenic follicles are associated with more than one arteriole. In humans, but not in rats, arterioles of identical size may either be covered by a T cell sheath or run directly through a germinal centre or through a corona. Thus, the term "central arteriole", indicating that an arteriole is located centrally in the PALS, is extremely difficult to apply in humans. It may, in fact, only be applicable to the larger vessels. Three-dimensional reconstructions of the course of intrasplenic arteries and arterioles are required to solve this problem.

It is not astonishing that various models have been proposed for the blood supply of human splenic follicles (Weidenreich 1901; Jäger 1929; Snook 1950; Snook 1975; Schmidt et al. 1988). We have observed that the blood supply of a germinal centre may be provided by an arteriole, which branches from the central arteriole, crosses the PALS and gives off several capillaries within the secondary follicle. In addition, other arterioles leave the central arteriole at the periphery of the follicle and curve around the follicle in the perifollicular region. However, there must be a whole variety of alternative patterns of blood supply to the follicles, as is evident from the literature.

The direction of blood flow in the sialoadhesin-positive sheathed capillaries is especially enigmatic. It is unequivocal that these vessels have to be attributed to the PFZ. As the entire course of a sheathed capillary is not observed within one section, the scheme depicted in Fig. 8 is hypothetical. We speculate that most of the sheathed capillaries arise from unsheathed small arterioles curving around the follicle or coming from the red pulp. Sheathed and also unsheathed capillaries approaching the follicle from the periphery may then open into the spaces of the PFZ. These spaces may additionally be supplied from the opposite direction by capillaries of the primary or secondary follicle centre.

The PFZ provokes additional speculations. Why is this region morphologically different from the remainder of the red pulp? It appears that certain white blood cells such as granulocytes and monocytes, and sometimes also B cells and sialoadhesin-positive macrophages, have a special affinity for some unknown attractive structure at this location. The density of these cells is clearly much higher than that found in the blood and thus perifollicular "hemorrhage" is not a good explanation for the phenomenon. However, the number of leukocytes in the perifollicular zone is variable among the specimens. Thus, could the PFZ represent the human functional equivalent of the rodent marginal sinus with respect to lymphocyte recirculation? Could it be a compartment where all types of leukocytes have to decide whether to continue migration into the white or into the red pulp? When blood pours into the open spaces of the PFZ, the leukocytes might be arrested at this site. Granulocytes and monocytes might then migrate to the red pulp and the lymphocytes would be free to move into the white pulp. If this hypothesis is tenable, it remains to be explained how the blood is moved forward from the PFZ to the splenic cords, and finally, into the splenic sinuses.

8.1.1 Summary

The PFZ of human primary and secondary follicles contains blood-filled spaces not lined by endothelia. We develop the hypothesis that leukocytes are arrested in these spaces and move from there either into the white pulp or into the red pulp. The spaces of the PFZ may be supplied both by sheathed and unsheathed capillaries approaching the follicle from the periphery and also by capillaries coming from the follicle centre. The PFZ may thus represent the human functional equivalent of the marginal sinus found in rats and mice.

8.2 Blood Circulation in the Splenic Red Pulp: Subpopulations of Fibroblasts and their Role

Up to now a convincing model of the blood flow in the open part of the splenic red pulp circulation, i.e. in the splenic cords, does not exist. The spleen represents the only human organ where blood is normally contained in spaces not lined by endothelial cells (if the placenta is not considered). In the splenic cords, these blood-filled spaces do not have smooth tubular walls similar to blood vessels, but they are subdivided by a very irregular labyrinth of fibrous strands produced and ensheathed by fibroblasts.[1] These fibroblasts make direct contact with the blood and they must thus differ from fibroblasts at other sites. Interestingly, this fact has been ignored in most investigations of the spleen. A crucial question is how the erythrocytes in the splenic cords retain enough propulsive energy to move through the cord labyrinth and then, finally, cross the narrow slits in the walls of the sinuses. Erythrocyte washout kinetics from extracorporally perfused rat and cat spleens show the existence of three vascular compartments, a fast, an intermediate and a slow one. In rats, 0.9% of the splenic blood flow goes to the slow compartment, which is assumed to represent the splenic cords. This compartment contains 50% of the splenic erythrocytes with a half-time of 90 min

[1] We use the term "fibroblast" in a very broad sense to replace the older "reticulum cell" and to subsume all fibre-producing connective tissue cells of the splenic stroma. To simplify the nomenclature, we do not distinguish among fibroblasts and fibrocytes.

(Cilento et al. 1980). In cats, 10% of the flow goes to the slow and intermediate compartments containing 90% of the erythrocytes with a half-time of 17 min to 5 h (Levesque and Groom 1976; Groom 1987). Thus, although the data are different for individual species, it is apparent, that a large number of erythrocytes spends quite a long time in a splenic compartment with low perfusion. There are, however, only very few investigations on erythrocyte kinetics in vivo. Erythrocyte transit times have been determined to be in the range of 30 s to 1 min for the entire spleen in humans (Jandl and Aster 1967), but nothing is known about transit times through the open part of the circulation. Because of the very incomplete knowledge of splenic microcirculation in man, it cannot be excluded that there are special mechanisms promoting blood flow through the splenic cords. In this context, contractile cells may be of importance.

We thus hypothesise that the fibroblasts in the human spleen fulfil functions different from fibroblasts in other organs. One of these functions may be a more pronounced contractility. Several phenotypically distinct types of fibroblasts occur in the human spleen. There is a clear distinction among the fibroblasts in the white pulp and those present in the red pulp. These cells not only differ with respect to their cell-surface integrin adhesion molecules but also with respect to their production of extracellular matrix components (Liakka et al. 1991; Liakka and Autio-Harmainen 1992; van den Berg et al. 1993; Liakka 1994). In addition, their expression of microfilaments and intermediate filaments is also quite diverse. Those fibroblasts associated with the PALS and MZ show ultrastructural similarities to smooth muscle cells and express smooth muscle alpha-actin and cytokeratins 8 and 18 (Franke and Moll 1987; Toccanier-Pelte et al. 1987; Doglioni et al. 1990; Satoh et al. 1997). Although there are contradictory results about the distribution of extracellular matrix components in the adult human spleen (Liakka 1994; van den Berg et al. 1993), it is clear that this population of fibroblasts also expresses high amounts of certain extracellular matrix ligands and adhesion molecules like alpha 2 integrins. We have confirmed actin and cytokeratin staining of the cells and regard them as myofibroblasts. Their reaction pattern is much better revealed in cryostat, as compared to paraffin, sections. In cryostat sections, the fibroblasts in the splenic red pulp sometimes also stain for smooth muscle alpha-actin and for cytokeratins including cytokeratin 8, but in our hands these cells exhibit a reduced staining intensity compared to cells in the PALS or in the MZ. Obviously the myofibroblasts described by us in the latter two locations correspond to the "fibroblastic reticulum cells" detected by Müller-Hermelink et al. (1981). These authors give a precise description of the myofibroblasts in the human spleen and demonstrate that they are positive for smooth muscle myosin and alkaline phosphatase. A similar finding is also described by Pinkus et al. (1986) for smooth muscle myosin in human spleens.

Reticulum cells positive for cytokeratins 8 and 18 or smooth muscle actin have been reported to occur in other human lymphoid organs such as lymph nodes. We confirm this fact, but our impression is that the staining intensity of myofibroblasts in human tonsils and lymph nodes is sometimes weaker than in the spleen. An explanation for this phenomenon is given by Franke and Moll (1987) and Toccanier-Pelte et al. (1987). These authors show that the expression of cytokeratins and of smooth muscle alpha-actin in fibroblasts of human lymphoid organs may vary depending on local immune reactions. In fact, Gould et al. (1995) indicate that cytokeratin expression is extremely variable among single lymph node specimens, from almost absent to very strong. They assume that the production of cytokeratins in fibroblasts of lymph nodes may be somehow regulated by the surrounding T lymphocytes. With respect to the spleen, we and others show that the amount of smooth muscle actin expressed in the fibroblasts of the MZ and of the PALS is much higher than that in red pulp fibroblasts. We interpret this finding as indicating that differentiation of myofibroblasts is promoted by close contacts to migratory $CD4^{+}$ T cells (see below). Thrombomodulin expression has, however, so far not been described for human splenic myofibroblasts. The normal expression of this molecule by the myofibroblasts of the PALS and MZ is not totally surprising, because vascular smooth muscle cells have been reported to be able to synthesise thrombomodulin in case of endothelial injury (Fink et al. 1993).

Thus, there is a tendency towards differentiation of myofibroblasts in the spleen, which is, however, most pronounced in the periphery of the white pulp. In addition, smooth muscle alpha-actin-positive cells also occur within human splenic trabeculae and they form several layers of the splenic capsule. It is an unsolved enigma, whether the normal human spleen is contractile and whether myofibroblasts may contribute to contractility. In hypersplenic human spleens, it has been shown that a substantial contraction reducing the organ volume by up to 50% is provoked after injection of epinephrine in vivo (Schaffner et al. 1985). A comparable effect also occurs in normal subjects after exercise at the maximal predicted heart rate (Schaffner et al. 1985) or in professional breath-holding divers (Hurford et al. 1990). Similarly, after injection of radiolabelled red cells, supine exercise reduced splenic radioactivity to about 50% (Sandler et al. 1984). Epinephrine also causes a reduction of spleen size totalling up to 35% in normal persons (Knecht et al. 1988) and maximal exercise leads to a 40% reduction of splenic blood volume (Froelich et al. 1988). Splenic contraction after epinephrine has even been suggested as a parameter to measure involvement of the organ in Hodgkin's disease (Osadchaya et al. 1980). However, in spite of all these results, it cannot be concluded that "contraction" is an active mechanism involving the entire splenic parenchyma. In fact, it may be just a passive consequence of arterial narrowing and reduced perfusion. Sympathetic nerve fibres are associated with arteries and arterioles in the human spleen, but sympathetic innervation of the splenic white or red pulp, trabeculae or capsule has not been described in humans (Heusermann 1988). However, the innervation of the human spleen has not been thoroughly investigated up to now. It is even controversial whether a cholinergic innervation of this organ exists (Kudoh et al. 1979). At the present state of knowledge, it cannot be excluded that the contractility of splenic myofibroblasts, minimal as it may be, could somehow contribute to the propulsion of erythrocytes through the labyrinth of the splenic red pulp cords. How the white pulp myofibroblasts fit into this picture is not clear.

It is also uncertain in rats whether splenic contraction is of functional importance. In rat spleens, it has been demonstrated that a major portion of red pulp fibroblasts exhibit ultrastructural features similar to smooth muscle cells (Saito 1977). Thus, these cells may be comparable to the weakly asm-1-positive red pulp fibroblasts in man. In the white pulp of rats, the distribution of fibroblast subpopulations is partially different from that in man. Strongly asm-1-positive myofibroblasts do not occur in the MZ, but only in the PALS and along the entire course of the marginal sinus. The myofibroblasts of the rat PALS appear to be phenotypically similar to those in man, at least with respect to smooth muscle alpha-actin expression. Interestingly, fibroblasts in the PALS were shown to exhibit a special phenotype different from that of other fibroblast populations in rat spleens (Van den Berg et al. 1989).

The adrenergic innervation of the rat spleen has been investigated in detail. There is an intricate association of nerve fibres and immunocompetent cells in the PALS and in the follicles (Felten et al. 1987; Felten and Olschowka 1987). This phenomenon is interpreted as an example for the interaction of the autonomous nervous system with the immune system in lymphatic organs. Innervation of the myofibroblasts in the PALS may, however, be another reason for the prevalence of sympathetic nerve fibres in the white pulp.

8.2.1
Summary

It is enigmatic how the large number of erythrocytes contained in the splenic cords is finally moved into the splenic sinuses to leave the organ. Contraction of the human spleen after sympathetic stimulation has been described, but this phenomenon may be a consequence of decreased perfusion. We hypothesise that there is a special propensity of splenic fibroblasts towards differentiation to myofibroblasts, which may support propulsion of erythrocytes on a microscopic scale. The most conspicuous myofibroblasts are present in the human oPALS and in the MZ. These cells have a peculiar phenotype expressing smooth muscle alpha actin, cytokeratins 8 or 18 and thrombomodulin. We speculate that an intimate contact with recirculating $CD4^+$ T lymphocytes is inducing expression of these molecules in white pulp fibroblasts.

8.3
Function of Sheathed Capillaries

The sheathed capillaries of the human spleen still represent enigmatic structures. Only very few reports have been concerned with these vessels (Schweigger-Seidel 1863; Solnitzky 1937; Brozman 1984; Buyssens et al.1984). It is not clear where the sheathed capillaries originate and where they end. Hueck (1927) and Jäger (1929) were the first to indicate that a large number of the sheathed vessels occur in the vicinity of follicles. Solnitzky (1937) and Buyssens et al. (1984) correctly diagnosed the sheath cells as "histiocytes" and correctly stated that the sheathed vessel is a capillary. Brozman (1984) demonstrated the sheathed capillaries at their correct location in humans. Buckley et al. (1991) characterised the sheath macrophages as being positive for CD14 and CD4, although they failed to attribute the sheathed capillaries to their correct microanatomical location. In general, the function and the predominant localisation of the sheaths appear to have been misinterpreted up to now. Buyssens et al. (1984) reported that lymphocytes occur within the capillary sheaths in humans. They speculated that the sheaths may represent structures of lymphocyte extravasation similar to the high endothelial venules in lymph nodes. We can, however, not confirm this observation. We agree with the assumption of Buyssens et al. (1984) that capillary sheaths are dynamic structures. Interestingly, mobility of sheath macrophages was postulated by Solnitzky (1937) and Van Krieken et al. (1988), although these authors did not present convincing evidence.

In fact, microanatomical peculiarities giving a clue to the function of capillary sheaths in humans do not exist. Thus, we can only add a further speculation derived from the following three findings:

1. Capillary sheaths occur in close vicinity to the B cell region of the white pulp.[1]
2. The sheath cells in the PFZ are macrophages expressing high amounts of sialoadhesin.
3. The sheaths dissolve under certain conditions and macrophages of the sheath phenotype distribute in the PFZ.

It has been shown in animals that the sheathed capillaries are rather leaky. In many species this fact enables sheath macrophages to sample particulate material and immune complexes from the blood (Espenes et al. 1995). We should thus like to imagine human sheath macrophages as taking up certain material, for example bacterial antigens, which cannot be collected at other locations. Our hypothesis is that the activated macrophages leave the sheaths during severe immune reactions to interact with cells in the PFZ or even with B lymphocytes of the outer MZ to promote B cell differentiation. Thus, the human sheath macrophages might correspond to the MZMϕ of rats.

8.3.1 Summary

Human spleens exhibit capillaries with macrophage sheaths in the PFZ. These sheaths are not encountered in rats or mice. In other species phylogenetically distant from humans, the capillary endothelium is leaky and capillary sheath macrophages sample immune complexes and particulate material from the blood. We speculate that the human sheath macrophages are able to leave their normal position on special occasions and that they may interact with immunocompetent lymphocytes, for example with B lymphocytes, in the PFZ to promote immune reactions.

8.4 Lymphocyte Migration in the Human Splenic White Pulp – A Hypothesis

The interpretation of the findings described in the preceding chapters suffers from the impossibility of conducting functional and kinetic in vivo experiments in humans. This fact has deterred many researchers from becoming occupied with the human spleen at all. In order not to leave the reader with a feeling of descriptive vagueness at the end of this review, we take the risk of formulating a hypothesis about the initial phase of lymphocyte recirculation in the human spleen.

We speculate that the human PFZ, with its open blood-filled spaces, represents the functional equivalent of the rodent marginal sinus and its surroundings. Thus, a part

[1] It is not clear as to whether additional sialoadhesin-negative macrophage sheaths may exist in the red pulp under special conditions.

of the blood circulating through the spleen is delivered to the perifollicular spaces, potentially by side-branches of the central arteries. From the perifollicular region, recirculating lymphocytes move into the white pulp while the remainder of the blood cells, such as erythrocytes, granulocytes and monocytes, enter the splenic cords and sinuses. Thus, lymphocytes do not have to pass an endothelial lining at the border of the white pulp. The smallest vessels arriving in the PFZ may possess open ends and may directly deliver their contents into the perifollicular spaces. These spaces are formed solely by fibroblasts and fibres without participation of endothelia. There may be some affinity of recirculating lymphocytes to the perifollicular fibroblasts or to extracellular matrix components produced by them. We assume that $CD4^+$ T lymphocytes migrate from the PFZ into the MZ. There, the majority of the $CD4^+$ T cells move along the MZ myofibroblasts (or their products) in the direction of the PALS. The myofibroblasts may also produce chemotactic mediators selectively attracting $CD4^+$ T lymphocytes. The peculiar phenotype of the MZ myofibroblasts may be the cause as well as the consequence of $CD4^+$ T cell passage. If our hypothesis is correct, initial contacts among $CD4^+$ recirculating T cell blasts and recirculating antigen-stimulated B lymphocytes may already occur in the PFZ or in the MZ of the spleen.

$CD8^+$ T cells obviously use this pathway only infrequently or with different kinetics. As mentioned in Sect. 4.1.2, $CD8^+$ cells clearly have a preference for the red pulp. A major proportion of these cells may, however, represent NK cells, because these cells also express the alpha chain of CD8. How those $CD8^+$ T cells which reside in the PALS reach their destination, remains unsolved. In contrast to rats, the amount of $CD8^+$ T cells in the PALS is extremely variable among different human specimens. Thus, $CD8^+$ T cells either use recirculation routes different from those of $CD4^+$ T cells or they fail to be arrested by the MZ myofibroblasts.

Recirculating IgD^+ B lymphocytes may also arrive in the PFZ and then enter the MZ. Thus, the population of perifollicular B cells may be rather dynamic. This would explain why the number of these cells varies among the PFZ of different specimens and of different follicles. The migration of recirculating IgD^+ B cells across the MZ into the corona would lead to mixing of these cells with sessile IgD^- B memory cells. The rather indistinct border between the corona and the MZ may be due to this phenomenon. Mixing of IgD^+ and IgD^- B cells would not occur to such an extent in rats because the recirculating B cells need not cross the entire MZ after leaving the marginal sinus. The small rim of B lymphocytes, which accompanies the oPALS in humans, appears to represent a very small and often interrupted equivalent of the rodent MZ. This is suggested by the fact that the row of B cells at the periphery of the PALS is in direct continuity with the outer part of the MZ of the follicles (Fig. 8). In addition, only very few strongly IgD^+ cells are found at the periphery of the PALS and even fewer IgD^+ cells occur inside. In fact, the density of these cells is not much different from that in the surrounding red pulp. However, in humans the PALS has no clear-cut outer border as there is no marginal sinus. Thus, the PALS and the equivalent of the MZ are difficult to distinguish. The question arises, whether the B cells at the periphery of the human PALS are recirculating cells carrying only minimal amounts of IgD and whether they can form extrafollicular foci of plasma cell differentiation. At present this problem cannot be tackled for lack of a method which unequivocally identifies human plasma cells in situ. In addition, as most of the specimens do not contain full-blown secondary follicles, we would not expect to find extrafollicular foci. There are specimens with a dramatically increased number of B cells present within the PALS, especially specimens from patients with gastrointestinal malignancies. Thus, the number of B cells migrating into the PALS may be variable, but this phenomenon needs more detailed investigation.

The spleen is the secondary lymphatic organ that can be most easily accessed for in situ studies in rodents. Thus, the spleen has been the organ of choice to demonstrate the effects of targeted gene deletions or transgenic expression of cytokines and chemokines on the immune system. Over the last few years, there has been an ever increasing number of publications investigating rodent spleens. Our hypothesis should caution researchers that there are clear-cut species differences in splenic microarchitecture. These differences may determine the locations of encounter and cooperation among immunocompetent cells, as well as their dependence on adhesion molecules or cytokine and chemokine gradients.

The disparities between rodent and human spleens are clearly visible, but they cannot be tested experimentally in vivo. Thus, we put forward our hypothesis to encourage further in situ and in vitro investigations on the human spleen. Analysis of cytokine or chemokine gene expression in situ, adhesion assays or experiments using isolated sheathed capillaries in cell culture may permit at least some advances in understanding splenic function in humans. It is our hope that such investigations will contribute to verifying or rejecting the assumptions mentioned above. More precise information about the microanatomy of human lymphatic organs will undoubtedly improve the clinical relevance of the results obtained in rodents using modern molecular biological techniques.

Acknowledgements. The authors gratefully acknowledge T. H. Schiebler's suggestions and comments. Biozol Diagnostica Vertrieb GmbH, Eching, Germany, and DAKO Diagnostica GmbH, Hamburg, Germany, kindly sponsored a part of the costs of colour publication. Our thanks are due to P.R. Crocker, Department of Biochemistry, University of Dundee, Scotland, for the generous gift of the sialoadhesin antiserum and the MAb HSN-1 as well as to M.R. Parwaresch and H.H. Wacker, Institute of Pathology, University of Kiel, Germany, for MAbs KiM1-P, KiM2 and KiM9R. We also gratefully acknowledge gifts of MAbs ED1, ED2 and ED3 by C.D. Dijkstra, Department of Cell Biology and Immunology, Vrije Universiteit, Amsterdam, The Netherlands, and the gift of MAb R73 by T. Hünig, Institute of Virology and Immunobiology, University of Würzburg, Germany. The scientific cooperation and reading of the manuscript by V. Grau, Institute of Anatomy and Cell Biology, University of Marburg, Germany, and the perfect technical assistance of B. Herbst were of essential importance for the success of the investigations. We thank C. Fiebiger, Institute of Anatomy and Cell Biology, University of Marburg, for skilfully illustrating Figs. 1 and 8.

9 Summary

This review aims to elucidate the species-specific microanatomical structure of human spleens by comparison with the well-known splenic compartments in rats. In addition, we try to correlate the individual splenic compartments with their immunological functions by reviewing experiments performed in rodents, and to a lesser extent, in man. The immunohistological results in rats and man were obtained in our laboratory. With respect to human spleens, they represent new and partially unpublished findings.

In rats, the splenic white pulp consists of three compartments, the periarteriolar lymphatic sheath (PALS), the follicles and the marginal zone. PALS and follicles are delimited from the red pulp by the broad marginal zone. The PALS is a compartment where T lymphocytes predominate. It also contains specialised myofibroblasts and antigen-presenting dendritic cells, the so-called interdigitating dendritic cells. The outer PALS represents a recirculation pathway for B lymphocytes. Primary follicles are B lymphocyte migration compartments, which may develop into secondary follicles during immune reactions. Full-blown secondary follicles consist of a germinal centre formed by a dark and a light zone and of a surrounding follicle corona (also called mantle zone). The corona represents the remainder of the primary follicle. Apart from proliferating and differentiating B lymphocytes, a germinal centre is populated by follicular dendritic cells and by tingible body macrophages. The marginal zone is a B lymphocyte compartment, which is predominantly inhabited by IgM^+ IgD^- B memory cells. An irregular blood vessel, the marginal sinus, separates the PALS and the follicles from the marginal zone. The marginal sinus is accompanied by a chain of special macrophages, the marginal metallophilic macrophages. In addition, the marginal zone contains a second scattered macrophage population, the marginal zone macrophages. Both macrophage populations strongly express sialoadhesin. The marginal zone additionally serves as a recirculation compartment for B and T lymphocytes.

In human spleens, the marginal zone is also composed of IgM^+IgD^- B memory cells. In contrast to rats, this compartment is almost exclusively present around the follicles. A human secondary follicle comprises up to four zones, the germinal centre, the corona, the marginal zone and the perifollicular zone. The perifollicular zone is not found in rats. It is a zone containing small blood-filled spaces at the periphery of the follicle, where erythrocytes, B lymphocytes, monocytes and granulocytes accumulate. Thus, the perifollicular zone may as well be regarded as a part of the splenic red pulp. Similar to the splenic red pulp cords, its blood-filled spaces are not lined by endothelial cells. A marginal sinus delimiting the PALS and the follicles from the marginal zone does not occur in humans. In addition, there are no marginal metallophilic macrophages. In consequence, the PALS and the follicles are not clearly separated from the marginal zone. Macrophages reside in the marginal zone, but in contrast to rats, these cells do not express sialoadhesin. Interestingly, strong expression of sialoadhesin is found in the macrophages of capillary sheaths, which occur in the perifollicular zone. In some individuals, scattered macrophages in the perifollicular zone strongly express sialoadhesin.

A special type of myofibroblasts is present in the human PALS. These cells continue into the marginal zone and divide it into an inner and an outer compartment. The myofibroblasts are positive for smooth muscle alpha-actin, cytokeratin 8 and/or 18 and thrombomodulin. In the marginal zone, migratory $CD4^+$ T lymphocytes are closely associated with these cells. Thus, similar to rodents, the human marginal zone is a recirculation compartment for B and T lymphocytes.

From the present results, we develop the hypothesis that lymphocyte recirculation in the human spleen partially differs from that in rats. We speculate that side branches of the central arterioles deliver blood into the spaces of the perifollicular zone. At this location, leukocytes are arrested for some time. Recirculating $CD4^+$ T lymphocytes and recirculating B cells then migrate across the marginal zone into the PALS or into the corona, respectively. Erythrocytes, granulocytes and monocytes enter the red pulp cords or sinuses. $CD4^+$ T lymphocytes use the myofibroblasts of the marginal zone as guiding structures on their way to the PALS. The majority of recirculating $CD8^+$ T lymphocytes either migrate with different kinetics, in different numbers or even take another route.

The spleen is an essential organ for combating invasion of the blood by encapsulated bacteria. Antibody production against bacterial capsular polysaccharides, which function as T-independent type 2 antigens, is correlated with B cells of the marginal zone. The role of marginal zone macrophages, which take up and store these antigens in rodents, is still controversial.

The immunological role of splenic compartments has been most thoroughly studied during antibody responses against protein antigens in rodents. These antigens necessitate an interaction of $CD4^+$ T helper cells with B cells to promote B cell differentiation towards antibody production. There are several consecutive phases of a T cell-dependent B cell reaction in the spleen.

First, $CD4^+$ helper T cells have to be stimulated by antigen-presenting class II MHC-positive dendritic cells in the PALS. The interdigitating dendritic cells are the most likely candidates for this function. Stimulated T cells then migrate to the outer PALS, where they encounter activated recirculating B cells. These B cells have bound and internalised their specific antigen with the help of their surface immunoglobulin. Similar to dendritic cells, B cells are able to present antigenic peptides to activated T cells. This process provokes help by the T cells, i.e. a second stimulatory signal is provided to the B lymphocytes.

In mice, B cells which have met their specific antigen and which have received a second T cell signal settle down in so-called extrafollicular foci at the outer border of the PALS to differentiate into antibody-secreting cells. These B cells are short-lived and the secreted antibodies are of low affinity. Circulating antibodies form a prerequisite for the formation of germinal centres. Antigen-antibody complexes are bound to the surface of follicular dendritic cells within follicles, where they persist for very long periods.

If recirculating B cells meet their specific antigen in the immune complexes on follicular dendritic cells in a primary follicle, the B cells are arrested and they start to proliferate. The proliferating B cells give rise to a germinal centre and antigen-specific

$CD4^+$ helper T cells are attracted to the site. The germinal centre develops a dark zone consisting of centroblasts and a light zone with centrocytes. Centroblasts hypermutate the antigen-binding portions of their immunoglobulin genes and they are prone to apoptosis. The surviving centroblasts differentiate into centrocytes and move into the light zone of the germinal centre. At this location, centrocytes are positively selected by contact with $CD4^+$ helper T cells and follicular dendritic cells. Before leaving the germinal centre they are programmed to differentiate either into B memory cells or into plasma cells.

Plasma cell precursors from germinal centres are distributed by the blood and settle in connective tissues to complete their final maturation. They represent long-lived cells secreting antibodies of high and maturing affinity. The most important site of long-term plasma cell immunoglobulin secretion is the bone marrow. B memory cells recirculate between blood and lymph before terminal differentiation to plasma cells. A special pathway leads to B memory cells which continue expressing IgM without IgD and which do not switch to other Ig isotypes. These cells migrate to the splenic marginal zone and settle there.

10 References

Ahmed R, Gray D (1996) Immunological memory and protective immunity: understanding their relation. Science 272: 54-6006

Alimzhanov MB, Kuprash DV, Kosco-Vilbois MH, Luz A, Turetskaya RL, Tarakhovsky A, Rajewski K, Nedospasov SA, Pfeffer K (1997) Abnormal development of secondary lymphoid tissues in lymphotoxin beta-deficient mice. Proc Natl Acad Sci USA 94: 9302–9307

Altschul R, Hummason FA (1947) Minimal vascular injection of the spleen. Anat Rec 97: 259–264

Amlot PL, Hayes AE (1985) Impaired human antibody response to the thymus-independent antigen, DNP-ficoll, after splenectomy. Lancet I:1008–1011

Ardavin C (1997) Thymic dendritic cells. Immunol Today 18: 350–361

Arpin C, Dechanet J, van Kooten C, Merville P, Grouard G, Briere F, Banchereau J, Liu Y-J (1995) Generation of memory B cells and plasma cells in vitro. Science 268: 720–722

Arpin C, Banchereau J, Liu Y-J (1997) Memory B cells are biased towards terminal differentiation: A strategy that may prevent repertoire freezing. J Exp Med 186: 931–940

Bachmann MF, Hoffmann Rohrer U, Kündig TM, Bürki K, Hengartner H, Zinkernagel RM (1993) The influence of antigen organization on B cell responsiveness. Science 262: 1448–1451

Bachmann MF, Kündig TM, Odermatt B, Hengartner H, Zinkernagel RM (1994) Free recirculation of memory B cells versus antigen-dependent differentiation to antibody-forming cells. J Immunol 153: 3386–3397

Bachmann MF, Odermatt B, Hengartner H, Zinkernagel RM (1996) Induction of long-lived germinal centers associated with persisting antigen after viral infection. J Exp Med 183: 2259–2269

Baggiolini M, Dewald B, Moser B (1997) Human chemokines: an update. Annu Rev Immunol 15: 675–705

Banchereau J, Steinman RM (1998) Dendritic cells and the control of immunity. Nature 392: 245–252

Baum M, Geitner T, Liesen H (1996) The role of the spleen in the leukocytosis of exercise: consequences for physiology and pathophysiology. Int J Sports Med 17: 604–607

Berek C, Milstein C (1988) The dynamic nature of the antibody repertoire. Immunol Rev 105: 5–26

Berek C (1992) The development of B cells and the B cell repertoire in the microenvironment of the germinal center. Immunol Rev 126:5–19

Billroth T (1861) Zur normalen und pathologischen Anatomie der menschlichen Milz. Virchows Arch Pathol Anat 20: 409–425

Bix M, Locksley RM (1995) Natural T cells. Cells that co-express NKRP-1 and TCR. J Immunol 155: 1020–1022

Bleul CC, Schultze JL, Springer TA (1998) B lymphocyte chemotaxis regulated in association with microanatomic localization, differentiation state, and B cell receptor engagement. J Exp Med 187: 753-762

Bordessoule D, Gaulard P, Mason DY (1990) Preferential localisation of human lymphocytes bearing $\gamma\delta$ T cell receptors to the red pulp of the spleen. J Clin Pathol 43: 461–464

Brelinska R, Pilgrim C (1983) Macrophages and interdigitating cells: their relationship to migrating lymphocytes in the white pulp of rat spleen. Cell Tissue Res 233: 671–688

Browning JL, Dougas I, Ngam-ek A, Bourdon PR, Ehrenfels BN, Miatkowski K, Zafari M, Yampaglia AM, Lawton P, Meier W, Benjamin CP, Hession C (1995) Characterization of surface lymphotoxin forms. J Immunol 154: 33–46

Brozman M (1984) Immuno- and enzyme histochemical studies on the sheathed capillaries in the marginal zone of the human spleen. Z Mikrosk-Anat Forsch 98: 284–292
Brozman M (1985a) Outer mantle zone of the follicle in the human spleen. Virchows Arch A (Pathol Anat) 407: 107–117
Brozman M (1985b) Anatomical pathways from the white to the red pulp in the human spleen. Acta Anat 121: 189–193
Buckley PJ, Dickson SA, Walker WS (1985) Human splenic sinusoidal lining cells express antigens associated with monocytes, macrophages, endothelial cells, and T lymphocytes. J Immunol 134: 2310–2315
Buckley PJ, Smith MR, Braverman MF, Dickson SA (1987) Human spleen contains phenotypic subsets of macrophages and dendritic cells that occupy discrete microanatomic locations. Am J Pathol 128: 505–520
Buckley PJ (1991) Phenotypic subpopulations of macrophages and dendritic cells in human spleen. Scanning Microsc 5: 147–158
Buiting AMJ, de Rover Z, Kraal G, van Rooijen N (1997) Humoral immune responses against particulate bacterial antigens are dependent on marginal metallophilic macrophages in the spleen. Scand J Immunol 43: 398–405
Buyssens N, Paulus G, Bourgeois N (1984) Ellipsoids in the human spleen. Virchows Arch A (Pathol Anat) 403: 27–40
Camacho SA, Kosco-Vilbois MH, Berek C (1998) The dynamic structure of the germinal center. Immunol Today 19: 511–514
Candinas D, Belliveau S, Koyamada N, MiyatakeT, Hechenleitner P, Mark W, Bach FH, Hancock WW (1996) T cell independence of macrophage and natural killer cell infiltration, cytokine production, and endothelial activation during delayed xenograft rejection. Transplantation 62: 1920–1927
Casamayor-Palleja M, Feuillard J, Ball J, Drew M, MacLennan ICM (1996) Centrocytes rapidly adopt a memory B cell phenotype on co-culture with autologous T cell-enriched preparations. Int Immunol 8: 737–744
Cella M, Engering A, Pinet V, Pieters J, Lanzavecchia A (1997) Inflammatory stimuli induce accumulation of MHC class II complexes on dendritic cells. Nature 388, 782–787
Chan EY-T, MacLennan ICM (1993) Only a small proportion of splenic B cells in adults are short-lived virgin cells. Eur J Immunol 23: 357–363
Chao D, MacPherson GG (1990) Analysis of thymus-independent type 2 antigen uptake by marginal zone macrophages in thin slices of viable lymphoid tissue in vitro. Eur J Immunol 20: 1451–1455
Chen L-T, Weiss L (1972) Electron microscopy of the red pulp of human spleen. Am J Anat 134: 425–458
Chen L-T and Weiss L (1973) The role of the sinus wall in the passage of erythrocytes through the spleen. Blood 41: 529–537
Chen X, Martin F, Forbush KA, Perlmutter RM, Kearney JF (1997) Evidence for selection of a population of multi-reactive B cells into the splenic marginal zone. Int Immunol 9: 27–41
Cilento EV, McCuskey RS, Reilly FD, Meineke HA (1980) Compartmental analysis of circulation of erythrocytes through the rat spleen. Am J Physiol 239: H272-H277
Claassen E, Kors N, Dijkstra CD, van Rooijen N (1986) Marginal zone of the spleen and the development of specific antibody-forming cells against thymus-dependent and thymus-independent type-2 antigens. Immunology 57: 399–403
Cook MC, Basten A, Fazekas de St. Groth B (1997) Outer periarteriolar lymphoid sheath arrest and subsequent differentiation of both naive and tolerant immunoglobulin transgenic B cells is determined by B cell receptor occupancy. J Exp Med 186: 631–643
Crocker PR, Gordon S (1989) Mouse macrophage hemagglutinin (sheep erythrocyte receptor) with specificity for sialylated glycoconjugates characterized by a monoclonal antibody. J Exp Med 169: 1333–1346
Crocker PR, Mucklow S, Bouckson V, McWilliam A, Willis AC, Gordon S, Milon G, Kelm S, Bradfield P (1994) Sialoadhesin, a macrophage sialic acid binding receptor for haematopoietic cells with 17 immunoglobulin-like domains. EMBO J 13: 4490–4503
Crowe PD, VanArsdale TL, Walter BN, Ware CF, Hession C, Ehrenfels B, Browning JL, Din WS, Goodwin RG, Smith CA (1994) A lymphotoxin-beta-specific receptor. Science 264: 707–710

Cyster JG, Hartley SB, Goodnow CC (1994) Competition for follicular niches excludes self-reactive cells from the recirculating B cell repertoire. Nature 371: 389–395
Cyster JG, Goodnow CC (1995) Pertussis toxin inhibits migration of B and T lymphocytes into splenic white pulp cords. J Exp Med 182: 581–586
Dijkstra CD, Döpp EA, Joling P, Kraal G (1985a) The heterogeneity of mononuclear phagocytes in lymphoid organs: distinct macrophage subpopulations in the rat recognized by monoclonal antibodies. Immunology 54: 589–599
Dijkstra CD, van Vliet E, Döpp EA, van der Lelij AA, Kraal G (1985b) Marginal zone macrophages identified by a monoclonal antibody: characterization of immuno- and enzyme-histochemical properties and functional capacities. Immunolgy 55: 23–30
Doglioni C, Dell'Orto P, Zanetti G, Iuzzolino P, Coggi G, Viale G (1990) Cytokeratin-immunoreactive cells of human lymph nodes and spleen in normal and pathological conditions. An immunocytochemical study. Virchows Arch A (Pathol Anat) 416: 479–490
Drenckhahn D, Wagner J (1986) Stress fibres in the splenic sinus endothelium in situ: molecular structure, relationship to the extracellular matrix, and contractility. J Cell Biol 102: 1738–1747
Duijvestijn AM, van Goor H, Klatter F, Majoor GD, van Bussel E, van Breda-Vriesman PJC (1992) Antibodies defining rat endothelial cells: RECA-1, a pan-endothelial cell-specific monoclonal antibody. Lab Invest 66: 459–466
Dunn-Walters DK, Isaacson PG, Spencer J (1995) Analysis of mutations in immunoglobulin heavy chain variable region genes of microdissected marginal zone (MGZ) B cells suggests that the MGZ of human spleen is a reservoir of memory B cells. J Exp Med 182: 559–566
Du P. Heyns A, Badenhorst PN, Lötter MG, Pieters H, Wessels P (1985) Kinetics and mobilization from the spleen of Indium-111-labeled platelets during platelet apheresis. Transfusion 25: 215–218
Egelund Christensen B, Jönsson V, Matre R, Tönder O (1978) Traffic of T and B lymphocytes in the normal spleen. Scand J Haematol 20: 246–257
Espenes A, McLaren-Press C, Dannevig BH, Landsverk T (1995) Immune complex trapping in the splenic ellipsoids of the rainbow trout (Oncorhynchus mykiss). Cell Tissue Res 282: 41–48
Fang Y, Xu C, Fu Y-X, Holers M, Molina H (1998) Expression of complement receptors 1 and 2 on follicular dendritic cells is necessary for the generation of a strong antigen-specific IgG response. J Immunol 1160: 5273–5279
Farhi DC, Ashfaq R (1996) Splenic pathology after traumatic injury. Am J Clin Pathol 105: 474–478
Fehr T, Rickert RC, Odermatt B, Roes J, Rajewsky K, Hengartner H, Zinkernagel RM (1998) Antiviral protection and germinal center formation, but impaired B cell memory in the absence of CD19. J Exp Med 188: 145–155
Felten DL, Ackerman KD, Wiegand SJ, Felten SY (1987) Noradrenergic sympathetic innervation of the spleen: I. Nerve fibres associate with lymphocytes and macrophages in specific compartments of the splenic white pulp. J Neurosci Res 18: 28–36
Felten SY, Olschowka J (1987) Noradrenergic sympathetic innervation of the spleen: II. Tyrosine hydroxylase (TH)-positive nerve terminals form synaptic-like contacts on lymphocytes in the splenic white pulp. J Neurosci Res 18: 37–48
Ferguson SE, Han S, Kelsoe G, Thompson CB (1996) CD28 is required for germinal centre formation. J Immunol 156: 4576–4581
Fink LM, Eidt JF, Johnson K, Cook JM, Cook CD, Morser J, Marlar R, Collins CL, Schaefer R, Xie S-S, Hsu S-M, Hsu P-L (1993) Thrombomodulin activity and localization. Int J Dev Biol 37: 221–226
Förster R, Mattis AE, Kremmer E, Wolf E, Brem G, Lipp M (1996) A putative chemokine receptor, BLR1, directs B cell migration to defined lymphoid organs and specific anatomic compartments of the spleen. Cell 87: 1037–1047
Franke WW, Moll R (1987) Cytoskeletal components of lymphoid organs. I. Synthesis of cytokeratins 8 and 18 and desmin in subpopulations of extrafollicular reticulum cells of human lymph nodes, tonsils, and spleen. Differentiation 36: 145–163
Froelich JW, Strauss HW, Moore RH, McKusick KA (1988) Redistribution of visceral blood volume in upright exercise in healthy volunteers. J Nucl Med 29: 1714–1718
Fu Y-X, Huang G, Wang Y, Chaplin DD (1998) B lymphocytes induce the formation of follicular dendritic cell clusters in a lymphtoxin alpha-dependent fashion. J Exp Med 187: 1009–1018
Fujita T (1974) A scanning electron microscope study of the human spleen. Arch Histol Jap 37: 187–216

Fujita T, Kashimura M, Adachi K (1982) Scanning electron microscopy (SEM) studies of the spleen-normal and pathological. Scanning Electron Microscopy I: 435–444

Gearing AJH, Beckett P, Christodoulou M, Churchill M, Clements J, Davidson AH, Drummond AH, Galloway WA, Gilbert R, Gordon JL, Leber TM, Mangan M, Miller K, Nayee P, Owen K, Patel S, Thomas W, Wells G, Wood LM, Woolley K (1994) Processing of tumour necrosis factor-alpha precursor by metalloproteinases. Nature 370: 555–557

Germain RN, Castellino F, Han R, Reis e Sousa C, Romagnoli P, Sadegh-Nasseri S, Zhong G-M (1996) Processing and presentation of endocytically acquired protein antigens by MHC class II and class I molecules. Immunol Rev 151: 5–30

Geuze HJ (1998) The role of endosomes and lysosomes in MHC class II functioning. Immunol Today 19: 282–287

Giorno R (1984) Unusual structure of human splenic sinusoids revealed by monoclonal antibodies. Histochemistry 81: 505–507

Goldschneider I, McGregor DD (1968) Migration of lymphocytes and thymocytes in the rat. I The route of migration from blood to spleen and lymph nodes. J Exp Med 127: 155–168

Gonzalez M, Mackay F, Browning JL, Kosco-Vilbois MH, Noelle RJ (1998) The sequential role of lymphotoxin and B cells in the development of splenic follicles. J Exp Med 187: 997–1007

Gould VE, Bloom KJ, Franke WW, Warren WH, Moll R (1995) Increased numbers of cytokeratin-positive interstitial reticulum cells (CIRC) in reactive, inflammatory and neoplastic lymphadenopathies: hyperplasia or induced expression? Virchows Arch 425: 617–629

Gray D, MacLennan ICM, Bazin H, Khan M (1982) Migrant $\mu^+\delta^+$ and static $\mu^+\delta^-$ B lymphocyte subsets. Eur J Immunol 12: 564–569

Gray D, McConnell I, Kumararatne DS, MacLennan ICM, Humphrey J, Bazin H (1984a) Marginal zone B cells express CR1 and CR2 receptors. Eur J Immunol 14: 47–52

Gray D, Kumararatne DS, Lortan J, Khan M, MacLennan ICM (1984b) Relation of intrasplenic migration of marginal zone B cells to antigen localization on follicular dendritic cells. Immunology 52: 659–669

Gray D, Skarvall H (1988) B cell memory is short- lived in the absence of antigen. Nature 336: 70–73

Gray D, Dullforce P, Jainandunsing S (1994) Memory B cell development but not germinal center formation is impaired by in vivo blockade of CD40-CD40 ligand interaction. J Exp Med 180: 141–155

Grell M, Douni E, Wajant H, Löhden M, Clauss M, Maxeiner B, Georgopoulos S, Lesslauer W, Kollias G, Pfizenmaier K, Scheurich P (1995) The transmembrane form of tumor necrosis factor is the prime activating ligand of the 80 kDa tumor necrosis factor receptor. Cell 83: 793–802

Grewal IS, Flavell RA (1996) The role of CD40 ligand in costimulation and T cell activation. Immunol Rev 153: 85–106

Groeneveld PHP, Erich T, Kraal G (1985) In vivo effects of LPS on B lymphocyte subpopulations. Migration of marginal zone-lymphocytes and IgD-blast formation in the mouse spleen. Immunobiol 170: 402–411

Grogan TM, Jolley CS, Rangel CS (1983) Immunoarchitecture of the human spleen. Lymphology 16: 72–82

Grogan TM, Rangel CS, Richter LC, Wirt DP, Villar HV (1984) Further delineation of the immunoarchitecture of the human spleen. Lymphology 17: 61–68

Groom AG (1987) Microcirculation of the spleen: new concepts, new challenges. Microvasc Res 34: 269–289

Grouard G, Durand I, Filgueira L, Banchereau J, Liu Y-J (1996) Dendritic cells capable of stimulating T cells in germinal centres. Nature 384: 364–367

Gulbranson-Judge A, MacLennan ICM (1996) Sequential antigen-specific growth of T cells in the T zones and follicles in response to pigeon cytochrome c. Eur J Immunol 26: 1830–1837

Gunn MD, Ngo VN, Ansel KM, Ekland EH, Cyster JG, Williams LT (1998) A B cell-homing chemokine made in lymphoid follicles activates Burkitt's lymphoma receptor-1. Nature 391: 799–803

Han SH, Hathcock K, Zheng B, Kepler TB, Hodes R, Kelsoe G (1995) Cellular interaction in germinal centers. Roles of CD40 ligand and B7-2 in established germinal centers. J Immunol 155: 556–567

Heusermann U (1988) Morphology and function of the human spleen. Enzyme histochemical and electron microscopy studies of the splenic lymphatic vessels, nerves and connective tissue structures. Veröff Pathol 129: 1–165

Herrlinger R (1949) Neue funktionell-histologische Untersuchungen an der menschlichen Milz. Z Anat Entwicklungsgesch 114: 340–365
Hueck W (1927) Über den Bau der Lymphknötchen in der Milz. Verh Dt Pathol Ges 22: 238–242
Kudoh G, Hoshi K, Murakami T (1979) Fluorescence microscopic and enzyme histochemical studies of the innervation of the human spleen. Arch Histol Jap 42: 169–180
Ho F, Lortan JE, MacLennan ICM, Khan M (1986) Distinct short-lived and long-lived antibody-producing cell populations. Eur J Immunol 16: 1297–1301
Hosea SW, Burch CG, Brown EJ, Berg RA, Frank MM (1981) Impaired immune response of splenectomised patients to polyvalent pneumococcal vaccine. Lancet I: 804–807
Hünig T, Wallny HJ, Hartley JK, Lawetzky A, Tiefenthaler G (1989) A monoclonal antibody to a constant determinant of the rat T cell antigen receptor that induces T cell activation. J Exp Med 169: 73–86
Hsu S-M, Cossman J, Jaffe ES (1983) Lymphocyte subsets in normal human lymphoid tissues. Am J Clin Pathol 80: 21–30
Hsu S-M (1985) Phenotypic expression of B lymphocytes. III. Marginal zone B cells in the spleen are characterized by the expression of Tac and alkaline phosphatase. J Immunol 135: 123–130
Humphrey JH (1981) Tolerogenic or immunogenic activity of hapten-conjugated polysaccharides correlated with cellular localization. Eur J Immunol 11: 212–220
Humphrey JH, Grennan D (1981) Different macrophage populations distinguished by means of fluorescent polysaccharides. Recognition and properties of marginal-zone macrophages. Eur J Immunol 11: 221–228
Humphrey JH (1985) Splenic macrophages: antigen presenting cells for TI-2 antigens. Immunol Lett 11: 149–152
Hurford WE, Hong, SK, Park YS, Ahn DW, Shiraki K, Mohri M, Zapol WM (1990) Splenic contraction during breath-hold diving in the Korean ama. J Appl Physiol 69: 932–936
Inaba K, Pack M, Sakuta H, Isdell F, Steinman RM (1997) High levels of a major histocompatibility complex II-self peptide complex on dendritic cells from the T cell areas of lymph nodes. J Exp Med 186, 665–672
Insel RA, Varade WS, Marin E (1994) Human splenic IgM immunoglobulin transcripts are mutated at high frequency. Mol Immunol 31: 383–392
Jacob J, Kassir R, Kelsoe G (1991a) In situ studies of the primary immune response to (4-hydroxy-3-nitrophenyl)acetyl. I. The architecture and dynamics of responding cell populations. J Exp Med 173: 1165–1175
Jacob J, Kelsoe G, Rajewsky K, Weiss U (1991b) Intraclonal generation of antibody mutants in germinal centres. Nature 354: 389–392
Jacob J, Kelsoe G (1992) In situ studies of the primary immune response to (4-hydroxy-3-nitrophenyl)acetyl. II. A common clonal origin for periarteriolar lymphoid sheath-associated foci and germinal centers. J Exp Med 176: 679–687
Jäger E (1929) Die Gefäßversorgung der Malpighischen Körperchen in der Milz. Zeitschr Zellforsch Mikr Anat 8: 578–601
Jandl JH, Aster RH (1967) Increased splenic pooling and the pathogenesis of hypersplenism. Am J Med Sci 253: 383–397
Kashimura M, Fujita T (1987) A scanning electron microscopy study of human spleen: relationship between the microcirculation and functions. Scanning Microscopy 1: 841–851
Klein U, Küppers R, Rajewsky K (1997) Evidence for a large compartment of IgM-expressing memory B cells in humans. Blood 89: 1288–1298
Knecht H, Jost R, Gmür J, Burger J, Fehr J (1988) Functional hyposplenia after allogeneic bone marrow transplantation is detected by epinephrine stimulation test and splenic ultrasonography. Eur J Haematol 41: 382–387
König R, Fleury S, Germain RN (1996) The structural basis of CD4-MHC class II interactions: co-receptor contributions to T cell receptor antigen recognition and oligomerization-dependent signal transduction. Curr Topics Microbiol Immunol 205: 19–46
Koni PA, Flavell RA (1998) A role for tumor necrosis factor receptor type 1 in gut-associated lymphoid tissue development: genetic evidence of synergism with lymphotoxin beta. J Exp Med 187: 1977–1983

Kosco-Vilbois MH, Gray D, Scheidegger, D, Julius M (1993) Follicular dendritic cells help resting B cells to become effective antigen-presenting cells: induction of B7/BB1 and upregulation of major histocompatibility complex class II molecules. J Exp Med 178: 2055–2066

Kosco-Vilbois MH, Zentgraf H, Gerdes J, Bonnefoy J-Y (1997) To "B" or not to "B" a germinal center? Immunol Today 5: 225–230

Kotani M, Matsuno K, Ezaki T (1986) Marginal zone bridging channels as a pathway for migrating macrophages from the red towards the white pulp in the rat spleen. Acta Anat 126: 193–198

Kotzé HF, du P Heyns A Wessels P, Pieters H, Badenhorst PN, Lötter MG (1986) Evidence that ^{111}In-labelled platelets pool in the spleen, but not in the liver of normal humans and baboons. Scand J Haematol 37: 259–264

Kraal G, Janse A (1986) Marginal metallophilic cells of the mouse spleen identified by a monoclonal antibody. Immunology 58: 665–669

Kraal G, Janse M, Claassen E (1988) Marginal metallophilic macrophages in the mouse spleen: effects of neonatal injections of MOMA-1 antibody on the humoral immune response. Immunology Lett 17: 139–144

Kraal G, Ter Hart H, Meelhuizen C, Venneker G, Claassen E (1989a) Marginal zone macrophages and their role in the immune response against T- independent type 2 antigens: modulation of the cells with specific antibody. Eur J Immunol 19: 675–680

Kraal G, Rodrigues H, Hoeben K, van Rooijen N (1989b) Lymphocyte migration in the spleen: the effect of macrophage elimination. Immunology 68: 227–232

Kraal G (1992) Cells in the marginal zone of the spleen. Int Rev Cytol 132: 31–74

Kraal G, Schornagel K, Streeter PR, Holzmann B, Butcher EC (1995) Expression of the mucosal vascular addressin, MAdCAM-1, on sinus-lining cells in the spleen. Am J Pathol 147: 763–771

Kroese FGM, Wubbena AS, Seijen HG, Nieuwenhuis P (1987) Germinal centers develop oligoclonally. Eur J Immunol 17: 1069–1072

Kronin V, Vremec D, Winkel K, Classon BJ, Miller RG, Mak TW, Shortman K, Süss G (1997) Are CD8$^+$ dendritic cells (DC) veto cells? The role of CD8 on DC in DC development and in the regulation of CD4 and CD8 T cell responses. Int Immunol 9: 1061–1064

Kumararatne DS, Gagnon RF, Smart Y (1980) Selective loss of large lymphocytes from the marginal zone of the white pulp in rat spleens following a single dose of cyclophosphamide. Immunology 40: 123–131

Kumararatne DS, Bazin H, MacLennan ICM (1981) Marginal zones: the major B cell compartment of rat spleens. Eur J Immunol 11: 858–864

Kumararatne DS, MacLennan ICM (1981) Cells of the marginal zone of the spleen are lymphocytes derived from recirculating precursors. Eur J Immunol 11: 865–869

Kündig TM, Bachmann MF, Oehen S, Hoffmann UW, Simard JJL, Kalberer CP, Pircher H, Ohashi PS, Hengartner H, Zinkernagel RM (1996a) On the role of antigen in maintaining cytotoxic T cell memory. Proc Natl Acad Sci USA 93: 9716–9723

Kündig TM, Bachmann MF, Ohashi PS, Pircher H, Hengartner H, Zinkernagel RM (1996b) On T cell memory: arguments for antigen dependence. Immunol Rev 150: 63–90

Küppers R, Zhao M, Hansmann M-L, Rajewsky K (1993) Tracing B cell development in human germinal centres by molecular analysis of single cells picked from histological sections. EMBO J 12: 4955–4967

Lane PJL, Gray D, Oldfield S, MacLennan ICM (1986) Differences in the recruitment of virgin B cells into antibody responses to thymus-dependent and thymus-independent type-2 antigens. Eur J Immunol 16: 1569–1575

Lanier LL, O'Fallon S, Somoza C, Philips J, Linsley PS, Okumura K, Ito D, Azuma M (1995) CD80 (B7) and CD86 (B70) provide similar co-stimulatory signals for T cell proliferation, cytokine production, and generation of CTL. J Immunol 154: 97–105

Lanier LL (1998) NK cell receptors. Ann Rev Immunol 16: 359–393

Lanzavecchia A (1998) Licence to kill. Nature 393: 413–414

Liu Y-J, Oldfield S, MacLennan ICM (1988) Memory B cells in T cell-dependent antibody responses colonize the splenic marginal zone. Eur J Immunol 18: 355–362

Lebecque S, de Bouteiller O, Arpin C, Banchereau J, Liu Y-J (1997) Germinal center founder cells display propensity for apoptosis before onset of somatic mutation. J Exp Med 185: 563–571

Lechler R, Aichinger G, Lightstone L (1996) The endogenous pathway of MHC class II antigen presentation. Immunol Rev 151: 51–79
Legler DF, Loetscher M, Stuber Roos R, Clark-Lewis I, Baggiolini M, Moser B (1998) B cell-attracting chemokine 1, a human CXC chemokine expressed in lymphoid tissues, selectively attracts B lymphocytes via BLR1/CXCR5. J Exp Med 187: 655–660
Le Hir M, Bluethmann H, Kosco-Vilbois M, Müller M, di Padova F, Moore M, Ryffel B, Eugster H-P (1996) Differentiation of follicular dendritic cells and full antibody responses require tumor necrosis factor receptor-1 signaling. J Exp Med 183: 2367–2372
Lenschow DJ, Walunas TL, Bluestone JA (1996) CD28/B7 system of T cell costimulation. Annu Rev Immunol 14: 233–258
Levesque MJ, Groom AC (1976) Washout kinetics of red cells and plasma from the spleen. Am J Physiol 231: 1665–1671
Li S, Satoh T, Korngold R, Huang Z (1998) CD4 dimerization and oligomerization: implications for T cell function and structure-based drug design. Immunol Today 19: 455–462
Liakka A Apaja-Sarkkinen M, Karttunen T, Autio-Harmainen H (1991) Distribution of laminin, and types IV and III collagen in fetal, infant and adult human spleens. Cell Tissue Res 263: 245–252
Liakka KA, Autio-Harmainen HI(1992) Distribution of the extracellular matrix proteins tenascin, fibronectin, and vitronectin in fetal, infant, and adult human spleens. J Histochem Cytochem 40: 1203–1210
Liakka KA (1994) The integrin subunits alpha 2, alpha 3, alpha 4, alpha 5, alpha 6, alpha V, beta 1 and beta 3 in fetal, infant and adult human spleen as detected by immunohistochemistry. Differentiation 56: 183–190
Lindhout E, Koopman G, Pals ST, de Groot C (1997) Triple check for antigen specificity of B cells during germinal centre reactions. Immunol Today 12, 573–577
Liu Y-J, Oldfield S, MacLennan ICM (1988) Memory B cells in T cell-dependent antibody responses colonize the splenic marginal zones. Eur J Immunol 18: 355–362
Liu Y-J, Zhang J, Lane PJL, Chan EY-T, MacLennan ICM (1991) Sites of specific B cell activation in primary and secondary responses to T cell-dependent and T cell-independent antigens. Eur J Immunol 21: 2951–2962
Liu Y-J (1997) Sites of B lymphocyte selection, activation, and tolerance in spleen. J Exp Med 186: 625–629
Liu Y-J, Banchereau J (1996) Mutant mice without B lymphocyte follicles. J Exp Med 184: 1207–1211
Liu Y-J, Banchereau J (1997) Regulation of B cell commitment to plasma cells or to memory B cells. Sem Immunol 9: 235–240
Llende M, Santiago-Delpin EA, Lavergne J (1986) Immunobiological consequences of splenectomy: a review. J Surg Res 40: 85–94
Lyons AB, Parish CD (1995) Are murine marginal-zone macrophages the splenic white pulp analog of high endothelial venules? Eur J Immunol 25: 3165–3172
MacDonald IC, Ragan DM, Schmidt EE, Groom AC (1987) Kinetics of red blood cell passage through interendothelial slits into venous sinuses in rat spleen, analyzed by in vivo microscopy. Microvasc Res 33: 118–134
Mackay CR (1997) What chemokine is that? Current Biology 7: R384-R386
MacLennan ICM, Liu Y-J (1991) Marginal zone B cells respond both to polysaccharide antigens and protein antigens. Res Immunol 142: 346–351
MacLennan ICM, Gulbranson-Judge A, Toellner K-M, Casamayor-Palleja M, Chan E, Sze D M-Y, Luther SA, Acha Orbea H (1997) The changing preference of T and B cells for partners as T dependent antibody responses develop. Immunol Rev 156: 53–66
Mandel TE, Phipps RP, Abbot A, TEW JG (1980) The follicular dendritic cell: long term antigen retention during immunity. Immunol Rev 53: 29–59
Martinez-Pomares L, Kosco-Vilbois M, Darley E, Tree P, Herren S, Bonnefoy J-Y, Gordon S (1996) Fc chimeric protein containing the cysteine-rich domain of the murine mannose receptor binds to macrophages from splenic marginal zone and lymph node subcapsular sinus and to germinal centers. J Exp Med 184: 1927–1937
Mason DY, Comans-Bitter WM, Cordell JL, Verhoeven M-AJ, van Dongen JJM (1990) Antibody L26 recognizes an intracellular epitope on the B cell associated CD20 antigen. Am J Pathol 136: 1215–1222

Matsumoto M, Lo SF, Carruthers CJL, Min J, Mariathasan S, Huang G, Plas DR, Martin SM, Geha RS, Nahm MN, Chaplin DD (1996a) Affinity maturation without germinal centres in lymphotoxin-alpha-deficient mice. Nature 382: 462–466

Matsumoto M, Mariathasan S, Nahm MH, Baranyay F, Peschon JJ, Chaplin DD (1996b) Role of lymphotoxin and the type I TNF receptor in the formation of germinal centers. Science 271: 1289–1291

Matsumoto M, Fu Y-X, Molina H, Huang G, Kim J, Thomas DA, Nahm MH, Chaplin DD (1997a) Distinct roles of lymphotoxin alpha and the type I tumor necrosis factor (TNF) receptor in the establishment of follicular dendritic cells from non-bone marrow-derived cells. J Exp Med 186: 1997–2004

Matsumoto M, Fu Y-X, Molina H, Chaplin DD (1997b) Lymphotoxin-alpha deficient and TNF receptor-I-deficient mice define developmental and functional characteristics of germinal centers. Immunol Rev 156: 137–144

Matsuno K, Ezaki T, Kotani M (1989) Splenic outer periarterial lymphoid sheath (PALS): an immunoproliferative microenvironment constituted by antigen-laden marginal metallophils and ED2-positive macrophages in the rat. Cell Tissue Res 257: 459–470

Matzinger P (1994) Tolerance, danger, and the extended family. Annu Rev Immunol 12: 991–1045

McMaster WR, Williams AF (1979) Identification of Ia glycoproteins in rat thymus and purification from rat spleen. Eur J Immunol 9: 426–433

Millikin PD (1969) The nodular white pulp of the human spleen. Arch Pathol 87: 247–258

Mitchell DA, Nair SK, Gilboa E (1998) Dendritic cell/macrophage precursors capture exogenous antigen for MHC class I presentation by dendritic cells. Eur J Immunol 28: 1923–1933

Mitchell J (1973) Lymphocyte circulation in the spleen. Marginal zone bridging channels and their possible role in cell traffic. Immunology 24: 93–107

Mond JJ, Lees A, Snapper CM (1995) T cell-independent antigens type 2. Annu Rev Immunol 13: 655–692

Mosier DE, Subbarao B (1982) Thymus-independent antigens: complexity of B lymphocyte activation revealed. Immunol Today 3: 217–222

Müller-Hermelink HK, v. Gaudecker B, Drenckhahn D, Jaworsky K, Feldmann C (1981) Fibroblastic and dendritic reticulum cells of lymphoid tissue. J Cancer Res Clin Oncol 101: 149–164

Ngo VN, Tang HL, Cyster JG (1998) Epstein-Barr virus-induced molecule 1 ligand chemokine is expressed by dendritic cells in lymphoid tissues and strongly attracts naive T cells and activated B cells. J Exp Med 188: 181–191

Nielsen HB, Secher NH, Kristensen JH, Christensen NJ, Espersen K, Pedersen BK (1997) Splenectomy impairs lymphocytosis during maximal exercise. Am J Physiol 272: R1847-R1852

Nieuwenhuis P, Ford WL (1976) Comparative migration of B and T lymphocytes in the rat spleen and lymph nodes. Cell Immunol 23: 254–267

Nossal GJV, Austin CM, Pye J, Mitchell J (1966) Antigens in immunity. XII. Antigen trapping in the spleen. Int Arch Allergy Appl Immunol 29: 368–383

Osadchaya TI, Vasilo NI, Baisogolov GD (1980) Diagnosis of splenic involvement in Hodgkin's disease by radionuclide evaluation of splenic contraction in response to adrenaline. J Nucl Med 21: 384–386

Pabst R (1988a) The spleen in lymphocyte migration. Immunol Today 9: 43–45

Pabst, R (1988b) The role of the spleen in lymphocyte migration. In: Husband AJ (ed) Migration and homing of lymphoid cells. CRC, Boca Raton, pp 63–84

Parums DV, Cordell JL, Micklem K, Heryet AR, Gatter KC, Mason DY (1990) JC70: a new monoclonal antibody that detects vascular endothelium associated antigen on routinely processed tissue sections. J Clin Pathol 43: 752–757

Pasparakis M, Alexopoulou L, Episkopou V, Kollias G (1996) Immune and inflammatory responses in TNF alpha-deficient mice: a critical requirement for TNF alpha in the formation of primary B cell follicles, follicular dendritic cell networks and germinal centers, and in the maturation of the humoral immune response. J Exp Med 184: 1397–1411

Pellas TC, Weiss L (1990) Deep splenic lymphatic vessels in the mouse: a route of splenic exit for recirculating lymphocytes. Am J Anat 187: 347–354

Peset Llopis, M-J, Harms G, Hardonk MJ, Timens W (1996) Human immune response to pneumococcal polysaccharides: complement-mediated localization preferentially on

CD21-positive splenic marginal zone B cells and follicular dendritic cells. J Allergy Clin Immunol 97: 1015–1024
Peters AM, Saverymuttu SH, Keshavarzian A, Bell RN, Lavender JP (1985) Splenic pooling of granulocytes. Clin Science 68: 283–289
Peters JH, Gieseler R, Thiele B, Steinbach F (1996) Dendritic cells: from ontogenetic orphans to myelomonocytic descendants. Immunol Today 17: 273–277
Peters M, Klonizakis I, Lavender JP, Lewis SM (1980) Use of ^{111}Indium-labelled platelets to measure spleen function. Br J Haematol 46: 587–593
Pettersen JC, Borgen DF, Graupner KC (1967) A morphological and histochemical study of the primary and secondary immune responses in the rat spleen. Am J Anat 121: 305–318
Pfeffer K, Mink K, Huffstadt U, Fütterer A (1997) The lymphotoxin β receptor (LTβ-R) controls the development of peripheral lymphoid organs. Immunobiology 197: 143
Pierre P, Turley SJ, Gatti E, Hull M, Meltzer J, Mirza A, Inaba K, Steinman RM, Mellman I (1997) Developmental regulation of MHC class II transport in mouse dendritic cells. Nature 388: 787–792
Pieters J (1997) MHC class II restricted antigen presentation. Curr Opinion Immunol 9: 89–96
Pinkus GS, Warhol MJ, O'Connor EM, Etheridge CL, Fujiwara K (1986) Immunohistochemical localization of smooth muscle myosin in human spleen, lymph node, and other lymphoid tissues. Am J Pathol 123: 440–453
Porgador A, Gilboa E (1995) Bone marrow-generated dendritic cells pulsed with class I-restricted peptide are potent inducers of cytotoxic T lymphocytes. J Exp Med 182: 255–260
Radzun HJ, Parwaresch MR (1983) Differential immunohistochemical resolution of the human mononuclear phagocyte system. Cell Immunol 82: 174–183
Radzun HJ, Hansmann M-L, Heidebrecht HJ, Bödewadt-Radzun S, Wacker HH, Kreipe H, Lumbeck H, Hernandez C, Kuhn C, Parwaresch MR (1991) Detection of a monocyte/macrophage differentiation antigen in routinely processed paraffin-embedded tissues by monoclonal antibody Ki-M1P. Lab Invest 65: 306–315
Randall TD, Heath AW, Santos-Argumedo L, Howard MC, Weissman IL, Lund FE (1998) Arrest of B lymphocyte terminal differentiation by CD40 signaling: mechanism for lack of antibody-secreting cells in germinal centers. Immunity 8: 733–742
Reis e Sousa C, Hieny S, Scharton-Kersten T, Jankovic D, Charest H, Germain RN, Sher A (1997) In vivo microbial stimulation induces rapid CD40 ligand-independent production of interleukin 12 by dendritic cells and their redistribution to T cell areas. J Exp Med 186: 1819–1829
Rennert PD, Browning JL, Hochman PS (1997) Selective disruption of lymphotoxin ligands reveals a novel set of mucosal lymph nodes and unique effects on lymph node cellular organization. Int Immunol 9: 1627–1639
Ridge P, Di Rosa F, Matzinger P (1998) A conditioned dendritic cell can be a temporal bridge between a $CD4^+$ T helper and a T killer cell. Nature 393: 474–478
Rot A (1993) Chemokines link the two steps of leukocyte adhesion to endothelium. The Immunologist 1: 145–149
Saito H (1977) Fine structure of the reticular cells in the rat spleen, with special reference to their fibro-muscular features. Arch Histol Jap 40: 333–345
Saitoh K, Kamiyama R, Hatakeyama S (1982) A scanning electron microscopic study of the boundary zone of the human spleen. Cell Tissue Res 222: 655–665
Sallusto F, Cella M, Danieli C, Lanzavecchia A (1995) Dendritic cells use macropinocytosis and the mannose receptor to concentrate macromolecules in the major histocompatibility complex class II compartment: down regulation by cytokines and bacterial products. J Exp Med 182: 389–400
Sandler MP, Kronenberg MW, Forman MB, Wolfe OH, Clanton JA, Partain CL (1984) Dynamic fluctuations in blood and spleen radioactivity: splenic contraction and relation to clinical radionuclide volume calculations. J Am Coll Cardiol 3: 1205–1211
Sasou S, Satodate R, Katsura S (1976) The marginal sinus in the perifollicular region of the rat spleen. Cell Tissue Res 172: 195–203
Sasou S, Satodate R, Masuda T, Takayama K (1986) Scanning electron microscopic features of spleen in the rat and human: a comparative study. Scanning Microscopy III: 1063–1069
Satodate R, Tanaka H, Sasou S, Sakuma T, Kaizuka H (1986) Scanning electron microscopical studies of the arterial terminals in the red pulp of the rat spleen. Anat Rec 215: 214–216

Satoh T, Takeda R, Oikawa H, Satodate R (1997) Immunohistochemical and structural characteristics of the reticular framework of the white pulp and marginal zone in the human spleen. Anat Rec 249: 486–494
Saunders D, Lucas K, Ismaili J, Wu L, Maraskovsky E, Dunn A, Shortman K (1996) Dendritic cell development in culture from thymic precursor cells in the absence of granulocyte/macrophage colony-stimulating factor. J Exp Med 184: 2185–2196
Schaffner A, Augustiny N, Otto RC, Fehr J (1985) The hypersplenic spleen. A contractile reservoir of granulocytes and platelets. Arch Intern Med 145: 651–654
Schaniel C, Pardali E, Sallusto F, Speletas M, Ruedl C, Shimizu T, Seidl T, Andersson J, Melchers F, Rolink AG, Sideras P (1998) Activated murine B lymphocytes and dendritic cells produce a novel CC chemokine which acts selectively on activated T cells. J Exp Med 188: 451–463
Schittek B, Rajewsky K (1990) Maintenance of B cell memory by long-lived cells generated from proliferating precursors. Nature 346: 749–751
Schmidt EE, MacDonald IC, Groom AC (1988) Microcirculatory pathways in normal human spleen, demonstrated by scanning electron microscopy of corrosion casts. Am J Anat 181: 253–266
Schmidt EE, MacDonald IC, Groom AC (1993) Comparative aspects of splenic microcirculatory pathways in mammals: the region bordering the white pulp. Scanning Microsc 7: 613–628
Schweigger-Seidel F (1863) Untersuchungen über die Milz. Virchows Arch Pathol Anat 27: 460–504
Shaw JHF, Print CG (1989) Postsplenectomy sepsis. Br J Surg 76: 1074–1081
Skalli O, Ropraz P, Trzeciak A, Benzonana G, Gillessen D, Gabbiani G (1986) A monoclonal antibody against alpha-smooth muscle actin: a new probe for smooth muscle differentiation. J Cell Biol 103: 2787–2796
Smith KGC, Hewitson TD, Nossal GJV, Tarlinton DM (1996) The phenotype and fate of the antibody-forming cells of the splenic foci. Eur J Immunol 26: 444–448
Smith ME, Ford WL (1983) The recirculating lymphocyte pool of the rat: a systematic description of the migratory behaviour of recirculating lymphocytes. Immunology 49: 83–94
Snapper CM, Mond JJ (1996) A model for induction of T cell-independent humoral immunity in response to polysaccharide antigens. J Immunol 157: 2229–2233
Snook T (1950) A comparative study of the vascular arrangements in mammalian spleens. Am J Anat 87: 31–77
Snook T (1964) Studies on the perifollicular region of the rat's spleen. Anat Rec 1148: 149–159
Snook T (1975) The origin of the follicular capillaries in the human spleen. Am J Anat 144: 113–117
Soares M, Havaux X, Van Beneden R, Van der Werf WJ, Bach FH, Bazin H, Latinne D (1995) Characterisation of rat anti-guinea pig circulating xenoreactive natural antibodies and secreting cells. Transpl Proc 27: 282–285
Solnitzky O (1937) The Schweigger-Seidel sheath (ellipsoid) of the spleen. Anat Rec 69: 55–75
Spencer J, Perry ME, Dunn-Walters DK (1998) Human marginal zone B cells. Immunol Today 19: 421–426
Spickett GP, Brandon MR, Mason DW, Williams AF, Woollett GR (1983) MRC Ox-22, a monoclonal antibody that labels a new subset of T lymphocytes and reacts with the high molecular weight form of the leukocyte-common antigen. J Exp Med 158: 795–810
Steiniger B, van der Meide PH (1993) High-dose interferon-gamma alters the distribution of B lymphocytes and macrophages in rat spleen and lymph nodes. Immunology 78: 461–467
Steiniger B, Barth P, Herbst B, Hartnell A, Crocker PR (1997) The species-specific structure of microanatomical compartments in the human spleen: strongly sialoadhesin-positive macrophages occur in the perifollicular zone, but not in the marginal zone. Immunology 92: 307–316
Steinman RM, Swanson J (1995) The endocytic activity of dendritic cells. J Exp Med 182: 283–288
Steinman RM, Pack M, Inaba K (1997) Dendritic cells in the T cell areas of lymphoid organs. Immunol Rev 156: 25–37
Stockinger B, Zal T, Zal A, Gray D (1996) B cells solicit their own help from T cells. J Exp Med 183: 891–899
Stuart AE, Warford A (1983) Staining of human splenic sinusoids and demonstration of unusual banded structures by monoclonal antisera. J Clin Pathol 36: 1176–1180
Süss G, Shortman K (1996) A subclass of dendritic cells kills CD4 T cells via Fas/Fas-ligand-induced apoptosis. J Exp Med 183: 1789–1796

Szakal AK, Kosco MH, Tew J (1988) A novel in vivo follicular dendritic cell-dependent iccosome-mediated mechanism for delivery of antigen to antigen-processing cells. J Immunol 140: 341–353
Szakal AK, Kosco MH, Tew JG (1989) Microanatomy of lymphoid tissue during humoral immune responses: structure function relationship. Annu Rev Immunol 7: 91–109
Szakal AK, Kapasi ZF, Masuda A, Tew JG (1992) Follicular dendritic cells in the alternative antigen transport pathway: microenvironment, cellular events, age and retrovirus related alterations. Sem Immunol 4: 257–265
Takahashi Y, Dutta PR, Cerasoli DM, Kelsoe G (1998) In situ studies of the primary immune response to (4-hydroxy-3-nitrophenyl)acetyl. V. Affinity maturation develops in two stages of clonal selection. J Exp Med 187: 885–895
Tanaka H, Takasaki S, Sakata A, Muroya T, Suzuki T, Ishikawa E (1984) Lymphocyte subsets in the white pulp of human spleen in normal and diseased cases. Acta Pathol Jpn 34: 251–270
Tew, JG, Kosco MH, Szakal AK (1989) The alternative antigen pathway. Immunol Today 10: 229–232
Tew JG, Wu J, Qin D, Helm S, Burton GF, Szakal AK (1997) Follicular dendritic cells and presentation of antigen and co-stimulatory signals to B cells. Immunol Rev 156: 39–52
Timens W, Boes A, Poppema S (1989) Human marginal zone B cells are not an activated B cell subset: strong expression of CD21 as a putative mediator for rapid B cell activation. Eur J Immunol 19: 2163–2166
Timens W, Poppema S (1985) Lymphocyte compartments in human spleen. An immunohistologic study in normal spleens and noninvolved spleens in Hodgkin's disease. Am J Pathol 120: 443–454
Timens W (1991) The human spleen and the immune system: not just another lymphoid organ. Res Immunol 142: 316–320
Tkachuk M, Bollinger S, Ryffel B, Pluschke G, Banks TA, Herren S, Gisler RH, Kosco-Vilbois MH (1998) Crucial role of tumor necrosis factor receptor1 expression on nonhematopoietic cells for B cell localization within the splenic white pulp. J Exp Med 187: 469–477
Toccanier-Pelte M-F, Skalli O, Kapanci Y, Gabbiani G (1987) Characterization of stromal cells with myoid features in lymph nodes and spleen in normal and pathologic conditions. Am J Pathol 129: 109–118
Toellner K-M, Gulbranson-Judge A, Taylor DR, Sze DM-Y, MacLennan ICM (1996) Immunoglobulin switch transcript production in vivo related to the site and time of antigen-specific B cell activation. J Exp Med 183: 2303–2312
Torres-Nagel N, Kraus E, Brown MH, Tiefenthaler G, Mitnacht R, Williams AF, Hünig T (1992) Differential thymus dependence of rat CD8 isoform expression. Eur J Immunol 22: 2841–2848
Van den Berg TK, Döpp EA, Brevé JJP, Kraal G, Dijkstra CD (1989) The heterogeneity of the reticulum of rat peripheral lymphoid organs identified by monoclonal antibodies. Eur J Immunol 19: 1747–1756
Van den Berg TK, Brevé JJP, Damoiseaux JGMC, Döpp E, Kelm S, Crocker PR, Dijkstra CD, Kraal G (1992) Sialoadhesin on macrophages: its identification as a lymphocyte adhesion molecule. J Exp Med 176: 647–655
Van den Berg TK, van der Ende M, Döpp EA, Kraal G, Dijkstra CD (1993) Localization of beta 1 integrins and their extracellular ligands in human lymphoid tissues. Am J Pathol 143: 1098–1110
Van den Eertwegh AJM, Boersma WJA, Claassen E (1992a) Immunological functions and in vivo cell-cell interactions of T cells in the spleen. Crit Rev Immunol 11: 337–380
Van den Eertwegh AJM, Laman JD, Schellekens MM, Boersma WJA, Claassen E (1992b) Complement-mediated follicular localization of T-independent type-2 antigens: the role of marginal zone macrophages revisited. Eur J Immunol 22: 719–726
Van der Valk P, van der Loo EM, Jansen J, Daha MR, Meijer CJLM (1984) Analysis of lymphoid and dendritic cells in human lymph node, tonsil and spleen. Virchows Arch B (Cell Pathol) 45: 169–185
Van Ewijk W and Nieuwenhuis P (1985) Compartments, domains and migration pathways of lymphoid cells in the splenic pulp. Experientia 4: 199–208
Van Gool SW, Vandenberghe P, de Boer M, Ceuppens JL (1996) CD80, CD86 and CD40 provide accessory signals in a multiple-step T cell activation model. Immunol Rev 153: 47–83
Van Kooten C, Banchereau J (1997) Functions of CD40 on B cells, dendritic cells and other cells. Curr Opinion Immunol 9: 330–337

Van Krieken JHJM, Te Velde J, Kleiverda K, Leenheers-Binnendijk L, Van de Velde CJH (1985) The human spleen; a histological study in splenectomy specimens embedded in methylmethacrylate. Histopathology 9: 571–585
Van Krieken JHJM, te Velde J (1986) Immunohistology of the human spleen: an inventory of the localization of lymphocyte subpopulations. Histopathology 10: 285–294
Van Krieken JHJM, te Velde J (1988) Normal histology of the human spleen. Am J Surg Pathol 12: 777–785
Van Krieken JHJM, von Schilling C, Kluin PM, Lennert K (1989) Splenic marginal zone lymphocytes and related cells in the lymph node: a morphologic and immunohistochemical study. Hum Pathol 20: 320–325
Van Rooijen N, van Nieuwmegen R, Kors N, Eikelenboom P (1984) The development of specific antibody-containing cells in the spleen of rabbits during the secondary immune response against free or liposome-associated albumin antigen. Anat Rec 208: 579–587
Van Rooijen N, Claassen E, Kraal G, Dijkstra CD (1989) Cytological basis of immune functions of the spleen. Immunocytochemical characterization of lymphoid and non-lymphoid cells involved in the "in situ" immune response. Progr Histochem Cytochem 19: 1–69
Van Rooijen N (1990) Direct intrafollicular differentiation of memory B cells into plasma cells. Immunol Today 11: 154–157
Van Vliet E, Melis M, van Ewijk W (1985) Marginal zone macrophages in the mouse spleen identified by a monoclonal antibody. J Histochem Cytochem 33: 40–44
Veerman AJP, van Ewijk W (1975) White pulp compartments in the spleen of rats and mice. A light and electron microscopic study of lymphoid and non-lymphoid cell types in T and B areas. Cell Tissue Res 156: 417–441
Vremec D, Zorbas M, Scollay R, Saunders DJ, Ardavin CF, Wu L, Shortman K (1992) The surface phenotype of dendritic cells purified from mouse thymus and spleen: investigation of the CD8 expression by a subpopulation of dendritic cells. J Exp Med 176: 47–58
Wacker HH (1994) Sinuswandzellen. Immunakzessorische Zellen des Lymphknotensinus. Progr. Pathol. 143
Ware CF, VanArsdale TL, Crowe PD, Browning JL (1995) The ligands and receptors of the lymphotoxin system. Curr Topics Microbiol Immunol 198: 175–217
Weidenreich F (1901) Das Gefäßsystem der menschlichen Milz. Arch Mikrosk Anat 58: 247–376
Weiss L, Powell R, Schiffman FJ (1985) Terminating arterial vessels in red pulp of human spleen: a transmission electron microscopic study. Experientia 41: 233–242
Willführ KU, Westermann J, Pabst R (1990) Absolute numbers of lymphocyte subsets migrating through the compartments of the normal and transplanted rat spleen. Eur J Immunol 20: 903–911
Witte T, Wordelmann K, Schmidt RE (1989) Heterogeneity of human natural killer cells in the spleen. Immunology 69: 166–170
Woollett GR, Barclay AN, Puklavec M, Williams AF (1985) Molecular and antigenic heterogeneity of the rat leukocyte-common antigen from thymocytes and T and B lymphocytes. Eur J Immunol 15: 168–173
Wu Y, Zhou Q, Zheng P, Liu Y (1998) CD28-independent induction of T helper cells and immunoglobulin class switches requires costimulation by the heat-stable antigen. J Exp Med 187: 1151–1156
Yamamoto K, Arimasa N, Yamamoto T, Tokuyama K, Kobayashi T, Itoshima T (1979) Scanning electron microscopy of the perimarginal cavernous sinus plexus of the human spleen. Scanning Electron Microsc III: 763–768
Yoshida K, van den Berg TK, Dijkstra CD (1994) The functional state of follicular dendritic cells in severe combined immunodeficiency (SCID) mice: role of the lymphocytes. Eur J Immunol 24: 464–468
Yoshida K, Kaji M, Takahashi T, van den Berg TK, Dijkstra CD (1995) Host origin of follicular dendritic cells induced in the spleen of SCID mice after transfer of allogeneic lymphocytes. Immunology 84: 117–126
Zamoyska R (1998) CD4 and CD8: modulators of T cell receptor recognition of antigen and of immune responses? Curr Opinion Immunol 10: 82–87

Zhang J, Liu Y-J, MacLennan ICM, Gray D, Lane PJL (1988) B cell memory to thymus-independent antigens type 1 and type 2: the role of lipopolysaccharide in B memory induction. Eur J Immunol 18: 1417–1424

Subject Index

GPSR Compliance

The European Union's (EU) General Product Safety Regulation (GPSR) is a set of rules that requires consumer products to be safe and our obligations to ensure this.

If you have any concerns about our products, you can contact us on ProductSafety@springernature.com

In case Publisher is established outside the EU, the EU authorized representative is:

Springer Nature Customer Service Center GmbH
Europaplatz 3
69115 Heidelberg, Germany

Batch number: 10761599

Printed by Printforce, the Netherlands